The Molecular and Hormonal Basis of
Plant-growth Regulation

The Molecular and Hormonal Basis
of Plant-growth Regulation

BY

YA'ACOV LESHEM

Department of Life Sciences, Bar-Ilan University
Ramat-Gan, Israel

ב'ה

PERGAMON PRESS

OXFORD · NEW YORK · TORONTO
SYDNEY · BRAUNSCHWEIG

Pergamon Press Ltd., Headington Hill Hall, Oxford
Pergamon Press Inc., Maxwell House, Fairview Park, Elmsford,
New York 10523
Pergamon of Canada Ltd., 207 Queen's Quay West, Toronto 1
Pergamon Press (Aust.) Pty. Ltd., 19a Boundary Street,
Rushcutters Bay, N.S.W. 2011, Australia
Vieweg & Sohn GmbH, Burgplatz 1, Braunschweig

First English Edition 1973
Translated and revised from the original Hebrew edition
published by Dvir Co., Tel Aviv, Israel, 1970

Library of Congress Cataloging in Publication Data
Leshem, Y
The molecular and hormonal basis of plant-growth regulation.
Translation of ha-Basis ha-molekulari veha-hormonali.
Includes bibliographies.
1. Hormones (Plants) 2. Molecular biology.
I. Title.
QK731.L4613 1973   581.3′1   73–6802
ISBN   0 08 017649 6

Printed in Great Britain by A. Wheaton & Co., Exeter

To Hasia and the crew

Contents

vii

Acknowledgements

I WISH to thank Prof. A. Halevi of the Faculty of Agriculture, The Hebrew University of Jerusalem, for reading the manuscript and for his helpful suggestions. Dr. D. Atsmon of the Plant Genetics Section and Prof. M. Revel of the Biochemistry Section of the Weizman Institute in Rehovoth provided kind advice on the subject-matter dealt with in Chapters 4 and 15. Numerous scientists the world over, whose names and institutional affiliations are mentioned in the text, have kindly provided illustrative material. I am also deeply indebted to the Bar-Ilan Research Council for encouragement and support during the writing and production of this text.

Rehovoth   YA'ACOV LESHEM
                                                    Sivan 5732

xi

Introduction

ALBERT EINSTEIN once said: "I believe in perfect laws in a world of existing things, in so far as they are real, which I try to understand with wild speculation." In this account of the molecular and hormonal basis of plant-growth regulation, the author has attempted to describe the existing knowledge and to distinguish clearly laws and speculation.

In the first part of his book the author describes the building stones that form the basis of present-day biology. The author inquires into the chemical and physical nature of the nucleic acids, considers how these can be a source of information which determine the character and, in part, the activities of the cell, and fit these into general biological theory.

The second part of the book deals with the modern aspects of hormone action introducing the reader to the growth-regulatory hormones existing in most higher plants. The molecular aspects of hormonal control have interested scientists for a relatively long time, but only recently with the development of modern technology has serious and accurate work been possible. The author being himself actively engaged in this field could critically and intelligently review this interesting field of work.

The briefness of this book, without loss of precision and of high scientific standard, makes it valuable for young scholars and biology teachers and provides them with a bird's-eye view of an interesting field. The bibliography after each chapter, arranged in a fashion that tells the reader the aim and scope of each quoted paper, will stimulate and permit him to widen the scope of his interest.

Although not being a classical textbook it is of value for all biologists, teachers and students who seek to widen their general knowledge and it is hoped that it will introduce the young plant physiologists into a field of great future.

PROF. B. LUNENFELD,
Professor of Biology,
Department of Life Sciences, Bar-Ilan University,
and Head, Institute of Endocrinology,
Tel-Hashomer Government Hospital, Israel

PART ONE

CHAPTER 1

Molecular Horizons

THE recent achievements in the understanding of basic biological processes which have paved the road towards a more thorough understanding of Life Sciences have been aided by the concept that all biological
entities are intrinsically similar and obey the same basic laws as regards
differentiation and growth. The biochemical approach, figuratively
speaking, has pulled the carpet from under the feet of classical genetics
and has produced a new science designated "Molecular Biology" which
in its turn has produced tenets of belief based on the "Central Dogma":

(chromosome) DNA replication $\xrightarrow{\text{transcription}}$ RNA $\xrightarrow{\text{translation}}$ protein

However, with the unravelling of mode of replication of certain RNA
viruses and phages a corollary has been added to the Dogma: the
sequence of the nucleic acids in the above statement may be reversed,
or RNA may completely replace DNA.

The combination of biological and physical scientific disciplines enabled Watson, Crick and Wilkins in 1953 to determine the structure of
DNA and this consequently led to the formulation of the genetic code
by Nirenberg, Matthei, Crick and Gamow in the early sixties. In 1957
Kornberg isolated an enzyme–DNA polymerase (now termed polymerase I) which was reported to participate in DNA replication. This
later led to the discovery of RNA polymerase participating in transcription (see Chapter 4) in 1960. Jacob and Monod in 1961 presented their
classic model of regulation of genetic information which now, with
certain important modifications, provides a unifying concept for the
understanding of the genetic processes in at least some of the biological Phyla. These are but a few of the several advances made by

3

molecular biology and in their wake practical applications in medicine, animal and plant sciences have followed even though many problems have yet to be solved.

To give examples of practical applications of molecular biology we may mention the attempt to transfer genetic information from one individual to another by means of transferring given and genetically mapped chromosomal fragments by what is termed DNA transduction. This is achieved by attaching these fragments to certain viruses which essentially are also genetic material and thereupon transferring the "enriched" virus to a new individual in which the whole particle is integrated into its DNA. By certain means the virus may subsequently be freed from the host DNA while leaving the chromosomal fragment of the donor with the recipient (this process will be discussed in detail in Chapter 9). Biochemical techniques are now sophisticated enough to enable the laboratory production of polynucleotides with base sequences corresponding to those of natural genes and Khorana's group in Wisconsin has achieved the *in vitro* synthesis of the gene for the alanine tRNA molecule. This is still a far call from the synthesis of a genetic entity enabling production of a complete protein or of a complete chromosome which may contain millions of nucleotide sequences, but nevertheless is an important step forwards towards therapy of genetic disorders caused by the inability to produce certain biologically active proteins.

Instances have been reported whereby not DNA but its transcriptional product, RNA, has been able to transfer genetic information. Several of the cases reported pertain to RNA mediated informational transfer from hormone-treated tissues; for instance, as induced by auxins in plants or by testosterone or oestrogens in mammals. RNA which was extracted from hormone-treated organs caused typical hormone-induced effects in non-hormone treated control organs.[3] Other RNA mediated effects concern transfer of information for protein enzyme synthesis, anti-body production and with certain reserve induction of memory. This subject will be elaborated upon in a later chapter. These techniques of DNA or RNA informational transfer present future possibilities of "genetic engineering" and while its applied utilization is still in its infancy it is probable that in the not too distant future some practical use may be made in this respect.

For generations horticulturalists have utilized the method of vegetative propagation of plant material, a practice as yet awaiting exploitation in animal husbandry. There is at least a theoretic possibility of non-sexual propagation of animal species and it has been claimed that *Homo sapiens* in this respect does not differ from other species. However, the possibility of "cloning" humans, even if at some future date this became practicable, would doubtlessly cause such great complexity of legal problems (e.g. parentage and inheritance) that this practice will be decided upon or rejected not by biologists but rather by legal and moral administrative authorities. A more feasible use of "test-tube" culture is that of certain tissues or differentiated organs which may provide specific genetic information, which by means of transduction, DNA or RNA transfer or other methods may be applied for therapeutic purposes.

In 1959 Steward at Cornell University succeeded in producing a perfectly normal carrot plantlet possessing root, shoot and flowers from tissue cultures of single vegetative cells growing in a nutrient medium containing growth substances, minerals, etc. (see Plate 1). This approach affords widespread agricultural use and in several instances has now been adopted commercially. Orchid growing normally necessitates an extensive period (several years) for the production of nursery stock and at present several nurseries, the world over, have developed commercial techniques whereby meristematic growing tips are cultured in a suitable and sterile medium, culture conditions causing the production of thousands of plantlets from a single tip. Once the plantlet stage has been reached the flasks are opened and the plantlets may be transferred to nursery beds and continue to develop normally. This technique saves 2 to 3 years of growth and has the further advantages of production of uniform material and, moreover, in the initial stages, is most economical in space. Plants produced from meristem culture are termed *mericlones*, and besides the practical use for orchids, successful experimentation has also been conducted on other "cash crops" such as chrysanthemums, carnations and bromelias, with special reference to the pineapple.

Steward's culture was of diploid (2n) cells stemming from lineage which at least in some previous stage underwent sexual fusion. Subsequent work has demonstrated the possibility of cloning haploid (n) cells as well, as achieved in the genus *Nicotiana*: the haploid tissue in

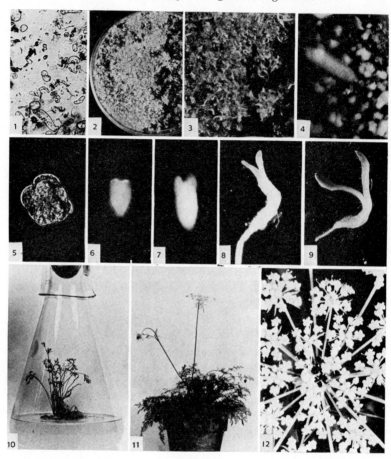

PLATE 1

The development of a carrot plant from individual root cells growing on nutrient medium under culture. In F. C. Steward's experiment the following stages can be distinguished: 1. Groups of individual cortical root cells from which cultures commenced. 2–5. Subsequent stages of tissue organization. Nos. 2 and 3 indicate development of groups of cells to form embryo-like organs (embryoids) which later produce plantlets. Nos. 4 and 5 are embryoids under magnification. 6. A normal carrot embryo. 7. An embryoid markedly resembling no. 6. 8–9. Later stages of development of the embryoid and the development of a plantlet. 10. Growth of plantlet on culture medium. 11. The plantlet has been potted and continues to grow normally. 12. Production of normal flowers on the potted plant. (Plate produced with the kind permission of F. C. Steward, Cornell University, U.S.A.)

this case is not necessarily somatic, since pollen grains under culture have also produced haploid plants.

The above experiments illustrate the concept of "totipotency", as coined by Bonner,[1] most convincingly. Totipotency implies that each and every cell contains the total sum of the genetic information required to produce the complete organism and under favourable conditions, such as those provided by Steward and outlined above, is capable of producing a complete new organism. This assumes the presence of a given amount of identical DNA in all plant cells independent of their differentiational state and specific functions. According to this hypothesis, the quantity of DNA per cell may vary only as a function of cell ploidy. Table 1 shows results of an experiment of Bonner's on maize (*Zea mays*) in which relative amounts of DNA in various cell types were compared.

TABLE 1. RELATIVE AMOUNTS OF DNA PER NUCLEUS IN VARIOUS CELL TYPES OF
MAIZE (*Zea mays*)
(According to Bonner)[1]

Cell type	Ploidy	Relative amount of DNA/nucleus
LEAF	2n	5·0
Root meristem—telophase	2n	5·0
Root meristem—prophase	4n	10·2
EMBRYO	2n	5·6
Seed—scutellum	2n	5·1
Seed—aleurone	3n	7·5
Microgamete	n	2·5

The data presented in the table indicate a clear correlation between amount of DNA and cell ploidy and that in all cells possessing similar ploidy, the amount of DNA is, within experimental error, equal. It is therefore inferred that development and differentiation of totipotential cells are dependent upon repression or depression mechanisms of genes which are defined regions of DNA. In most somatic cells of higher organisms it has been estimated that at the most only 5 per cent of the total DNA is utilized at a given period, different regions of the chromosome being turned on or off according to differentiation patterns. It is furthermore claimed that not all the information produced by active DNA in the nucleus reaches the cytoplasm, there to be translated, thus only a small proportion indeed of the totipotency of cells is finally expressed.

Disease control and especially viral disease control is being aided by molecular biology. In 1957 Lindenmann and Isaacs reported the existence of a substance named interferon which is produced in cells upon virus invasion and induces changes in the host cell, interfering with further multiplication of the invader while not hindering normal cell multiplication (see Chapter 11). Interferon which is formed by infection with an artificial virus—double-stranded RNA—has stemmed the spread of the viral eye infection herpetic keratoconjunctivitis in fowl. Research on the same lines is being conducted with the hope of combatting the common cold, the pathological cause of which is viral.

RECOMMENDED READING AND REFERENCES

1. BONNER, J., Development in *Plant Biochemistry*, pp. 850–72, ed. J. BONNER and J. E. VARNER, Academic Press, 1965. A semi-philosophical treatise on the subject of development, differentiation, growth and death.
2. PARK, J. H. and BARON, S., Herpetic keratoconjunctivitis: therapy with synthetic double-stranded RNA. *Science*, **162**, 811–13 (1968). A report on a series of experiments in rabbits in which viral eye disease was prevented or cured by means of synthetic RNA which induced interferon production.
3. SEGAL, S. J., *Control Mechanisms in Developmental Processes*, ed. LOCKE, M., Academic Press, 1967. A review article citing several instances of nucleic acid mediated hormonal information transfer.
4. SINSHEIMER, R. L., The prospects for designed genetic change. *Am. Scient.*, pp. 134–42 (1969). An article discussing possibilities of "genetic engineering" utilizing viruses as a means of gene transfer.
5. STEWARD, F. C., The control of growth in plant cells. *Scient. Am.* **209**, 104–13 (Oct. 1963). A description of differentiation commencing with single cells and climaxed by production of a completely normal plant using methods of tissue culture.
6. TAYLOR, G. T., *The Biological Time Bomb*, Thames & Hudson, 1968. A popular book by a scientific correspondent who predicts drastic changes in mankind if and when molecular biology will be harnessed. Also available as a paperback.

CHAPTER 2

The Structural Units of DNA

THE chromosome of higher plants and animals comprises a double chain of deoxyribonucleic acid (DNA). According to the model suggested by Watson, Crick and Wilkins the two chains are intertwined and form a "double helix". A gene is at present defined as a certain segment of a DNA molecule possessing specific structure which upon being active may finally translate itself into one polypeptide chain. This gene concept differs somewhat from the earlier contention made by Beadle and Tatum,[1] of "one gene, one protein". The "cistron" now is seen to be comprised of those stretches of DNA which participate in the formation of final polypeptide protein subunits. The basic structural units of DNA are:

1. The purine bases—adenine and guanine; and the pyrimidine bases—thymine and cytosine (see Fig. 1).
2. The sugar deoxyribose which is joined to the nucleic bases at serial position 1 of the sugar.
3. Phosphate groups which link up the sugar units by a 3,5-linkage forming a 3-5-phospho-diester polymer (see Fig. 2).

When one of the bases combines with the sugar ribose or deoxyribose a *nucleoside* is formed and when in addition to the two former units a phosphate group is included the product is termed a *nucleotide* (see Fig. 3). Besides the four common bases mentioned above, in certain types of DNA such as phage (bacterial virus) DNA, 5-hydroxymethylcytosine has been found, and in DNA of certain higher plants and animals respectively, 6 and 30 per cent of total nucleic base composition is methylcytosine. DNA of bacteria may contain small amounts of 6-methylaminopurine. These bases are termed minor bases and in

9

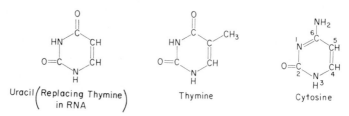

FIG. 1. The bases and sugars of nucleic acids.

Chapter 4 when discussing RNA, more will be said of them. The terminology used for some of the more common bases, their nucleosides and nucleotides is as follows:

The base	*Its nucleoside*	*Its nucleotide*
Adenine	adenosine	adenylic acid or adenosine monophosphate (AMP)
Guanine	guanosine	guanylic acid or guanosine monophosphate (GMP)
Cytosine	cytidine	cytidylic acid or cytidine monophosphate (CMP)
Thymine	thymidine	thymidylic acid or thymidine monophosphate (TMP)
Uracil	uridine	uridylic acid or uridine monophosphate (UMP)
Hypoxanthine	inosine	inosinic acid or inosine monophosphate (IMP)

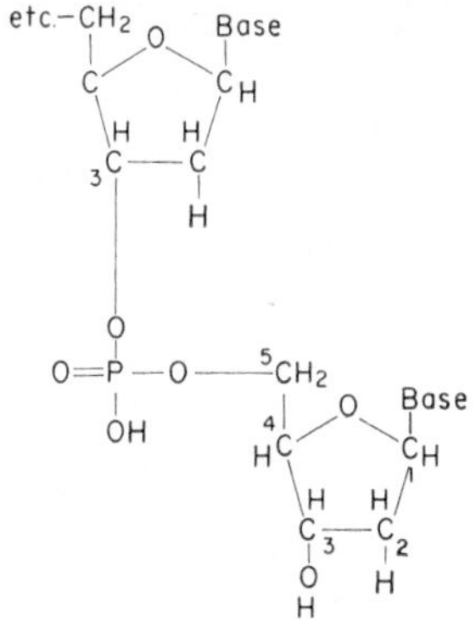

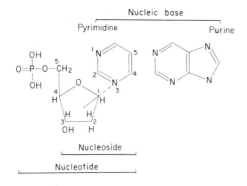

Fig. 2. The 3,5-diester bond.

Fig. 3. Nucleoside and nucleotide components.

The above-described units combined as shown in Fig. 4 form a single chain. The second chain resembles the first but its base sequence differs. There is very strong evidence that adenine of the one chain is paired with the thymine of the other, while guanine of one chain is invariably paired with cytosine (see Fig. 5). The bases are not covalently linked and are held together by hydrogen bonding. These bonds are comparatively weak—of the order of 3 to 7 kilocalories per mole as compared to the 50 to 100 kilocalories of regular covalent linkages.

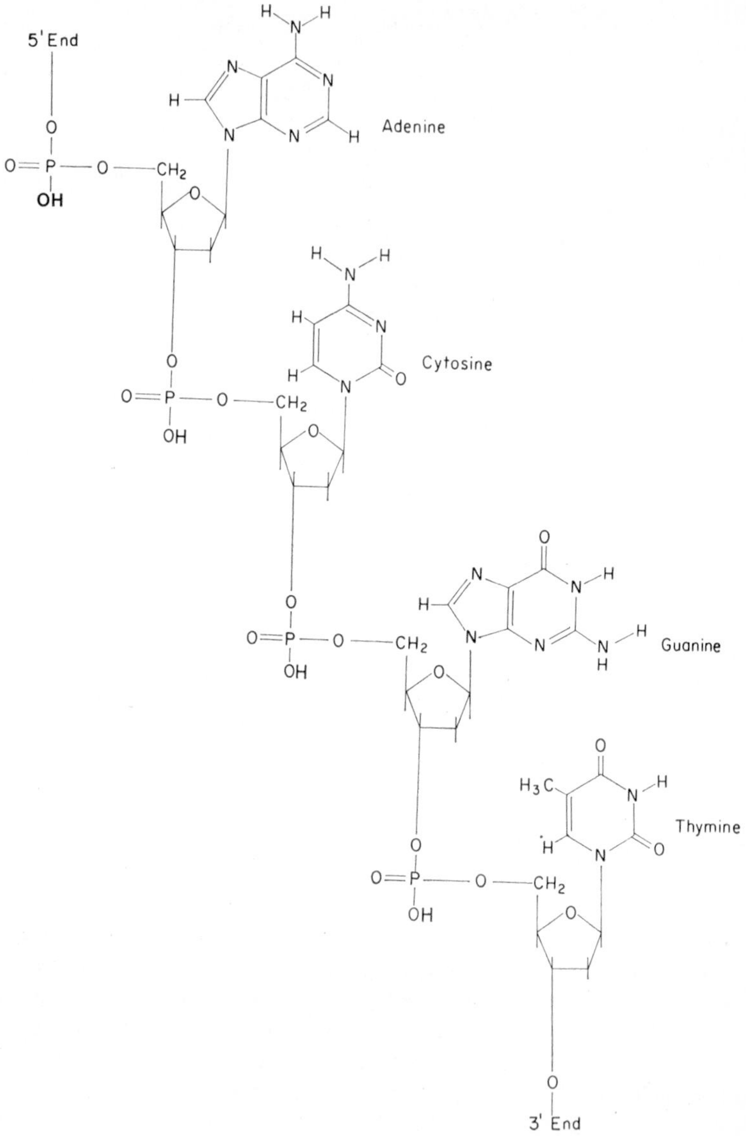

Fig. 4. DNA strand units. (After Watson[6].)

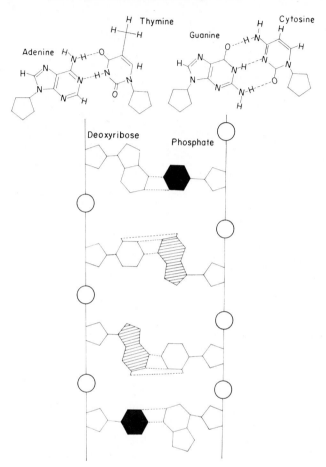

FIG. 5. *Above:* Base pairing by hydrogen bonding according to Crick and Watson. *Below:* The arrangement of the subunits to form a double-stranded molecule. The dotted lines indicate hydrogen bonds.

The weak hydrogen bonds are readily broken and re-formed at normal physiological temperatures without necessitating enzyme intervention. The final form assumed by the double-stranded DNA is helical and schematic configuration is shown in Fig. 6 (right).

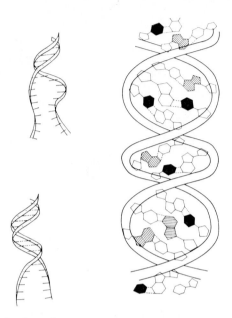

Fig. 6. *Right:* A schematic representation of the double helix. *Left:* Separation and reformation of the duplex. (After Watson[6].)

Despite the weak bonding the double-stranded structure is stable at physiological temperatures since many such links hold the units together, and even if several breaks occur at one end of the "duplex" molecule, the individual strands do not part company since usually a rapid reverse reaction takes place and the hydrogen bonds are re-formed (Fig. 6, left). Under certain circumstances as with high temperatures severance of hydrogen bonds becomes excessive and the strands separate completely, the phenomenon being termed DNA *melting*. Melting normally occurs only rarely and ten hydrogen-bonded nucleotide pairs are considered sufficient to stabilize the duplex.

DNA configuration

According to the model of Watson and Crick the two strands are complementary but are of opposite polarity, i.e. the free 5' end of the one strand corresponds to the 3′ end of the other. The nucleic bases are stacked at right angles to the longitudinal axis of the molecule and the helix is right-handed and a complete helical turn is described for every 34 Å. The distance between each base is approximately 3·4 Å and it may thus be calculated that each helical turn contains 34·0/3·4 = 10 complementary base pairs. The whole molecule may be visualized as a stack of these bases about which twine twisted ladders of pentose sugars and phosphate groups, the distance between the latter being 7·1 Å. The diameter of the molecules has been reported to be approximately 20 Å. These dimensions which were initially calculated theoretically by Watson and Crick were subsequently verified experimentally by X-ray crystallography by M. H. F. Wilkins and Rosemary Franklin.

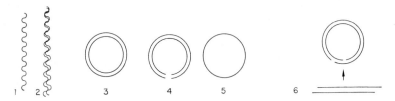

FIG. 7. Various DNA shapes. 1. Single linear strand. 2. Double linear strand. 3. Circular-double and closed. 4. Circular-double and open. 5. Circular-single. 6. Linear-circular with "sticky ends".

The length of the DNA molecule varies with species: especially high values have been reported, for the DNA of Chinese hamsters—1800 microns in the coiled double-helical state. In previous years reported lengths were considerably less for most investigated species since methods of extraction formerly used usually caused breakages in the DNA molecule. Upon uncoiling the strands the total length is much greater; for example, the average uncoiled single chromosomal length in mammals is 6 cm as compared with several millimetres in bacteria while in salamanders chromosomal lengths of 1 metre have been documented. From these figures it is apparent that the number of base pairs per chromosome is several billions.

Concerning the shape of the DNA molecule: it is not always linear but may be circular as in the bacterium *Escherichia coli* or the polyoma virus. The circular type of DNA is more resistant to enzymatic attack by a certain class of enzymes, the exonucleases, which could destroy the biological function of the molecule. Organisms in which the unit is not a classical double helix but rather a single-stranded DNA molecule are also known, e.g. phage $\phi \times 174$ (see Fig. 7). In organisms such as these it follows that the proportion of adenine has no relationship to that of thymine and neither is the amount of guanine dependent on that of cytosine. The circular form of DNA may be single or double stranded, "open" or "closed".

In linear types of DNA either single or double stranded, it often occurs that the ends "overlap", i.e. the ends of the chains with regard to base pairing are complementary to the heads of the chains and under certain conditions they may unite to form a circular molecule. Such ends are designated "sticky ends".

RECOMMENDED READING AND REFERENCES

1. BEADLE, G. W. and TATUM, E. L., Genetic control of biochemical reactions in *Neurospora. Proc. Nat. Acad. Sci.*, **27**, 499–506 (1941). A now classical paper which introduced a new gene concept.
2. BONNER, J., *The Molecular Biology of Development*, Clarendon Press, Oxford, 1965. A book dealing with general molecular biology problems with stress on methods and experimental proof of theories.
3. CRICK, F. H. C., The structure of hereditary material. *Scient. Am.* (Oct. 1954). A concise description of DNA stereochemistry. Written by Watson's co-recipient of the Nobel prize awarded in this field.
4. DuPRAW, E. J., *Cell and Molecular Biology*, Academic Press, N.Y., 1968. A fundamental textbook with thorough treatment of molecular biology as pertaining to plants, animals and bacteria.
5. WATSON, J. D., *Molecular Biology of the Gene*, Second Edition, Benjamin Inc., N.Y., 1970. A compulsory, lucidly written and well-illustrated treatise by the Nobel Prize winner.
6. WATSON, J. D., *The Double Helix*, Atheneum Publishers, N.Y., 1968. A highly dramatic, personal "behind the stage" account of how DNA structure was determined. Not a scientific text: available also as a paperback.
7. WATSON, J. D. and CRICK, F. H. C., Genetical implications of the structure of deoxyribonucleic acid. *Nature*, **171**, 964–77 (1953). The original report in which the authors presented the now accepted structure of genetic material.

CHAPTER 3

DNA Replication

DNA REPLICATION during mitotic division and tissue growth is still enigmatic and during the last decade the quest for the "True Replicase" has yielded several eligible enzymes. Be the enzyme mediator as it may the general scheme as outlined by A. Kornberg is represented in Fig. 8.

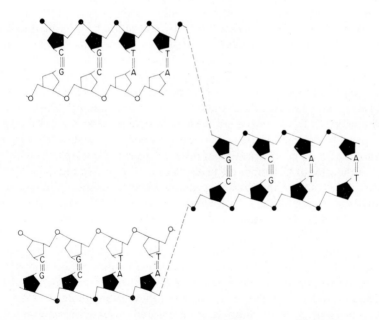

FIG. 8. Replication of the DNA duplex according to Kornberg.[4] Each parent strand leads to the formation of double-stranded daughter duplexes comprised of one parent strand and one new strand.

17

It can be seen that each parent chain forms a new chain, and the "daughter" duplexes comprise one new and one parent strand. The enzyme mediating this process was originally considered by Mitra and Kornberg[5] to be *DNA polymerase* or, as it is now more commonly known, *polymerase I*. It was claimed that this enzyme polymerizes the deoxynucleoside triphosphates (ATP, TTP, GTP and CTP), requires pre-existing DNA as a primer functioning in base pairing and also the presence of metallic ions such as Mg^{++} and Mn^{++}. The reaction takes place rapidly at the rate of about 1000 nucleotides per minute per molecule of enzyme, and direction of polymerization is from the 5' to the 3' end.

The concept of DNA polymerase I as the primary enzyme mediator of DNA replication has been reconsidered with the isolation by de Lucia and Cairns of an *E. coli* mutant which has less than 1 per cent of the normal (wild type) content of the Kornberg DNA polymerase but can grow and replicate its DNA normally. Further doubts were cast as to the relevance of polymerase I to DNA replication by the isolation by Knippers[3] in 1970 of a DNA synthesizing enzyme, called *polymerase II*, from a crude membrane fraction and independently in 1971 by T. Kornberg, the son of A. Kornberg, and Gefter[4] of a similar enzyme designated *polymerase III*. This enzyme was isolated and characterized from the polymerase-deficient mutant (the Cairns mutant). It was also assumed that DNA replication is likely to take place at the membrane of the cell. Polymerase III works best with double-stranded DNA as a template for its polymerization of deoxynucleoside triphosphates which as in the case of polymerase I takes place in 5' to a 3' direction. However, it differs from the Kornberg enzyme in being resistant to a specific antiserum which inhibits the Kornberg polymerase. The latter now has been relegated to the role of "repair", more of which shall be said presently.

The ruling out of polymerase I as the major replicase does not necessarily imply that this function is fulfilled by polymerase II or III. It has, however, led to the idea that *several* enzymes may exist in the cell, all of which are able to polymerize DNA, although most of them are not involved in duplication. A recent and novel approach has indicated that perhaps polymerization may be of nucleoside *mono* phosphates and not of triphosphates as previously believed, and for this reason the True Replicase has eluded detection—and meanwhile the quest continues.

The mode of replication has been shown to be "semi-conservative" whereby each new DNA duplex comprises one parent and one daughter strand (see Fig. 9). This was first shown by Taylor *et al.*[7] working on the broadbean *Vicia faba* and later demonstrated in bacteria as well. The "semi-conservative" model now is accepted and the "conservative" concept whereby each parent duplex leads to the formation of a completely new double-stranded duplex has been ruled out. This point has been elegantly demonstrated by growing bacteria in a culture medium in

FIG. 9. Conservative (right) and semi-conservative (left) DNA duplication.

which all the nitrogen supplied was the heavy N^{15} isotope and not the lighter normal N^{14} form. After several generations bacteria containing only N^{15} nitrogen were obtained and these were subsequently transferred to media containing N^{14}. The first-generation progeny of this transfer contained bacteria whose DNA was "hybrid" containing one light N^{14} and one heavy N^{15} chain, thus confirming the "semi-conservative" model.

Replication may proceed either "continuously" or "discontinuously". Sugimoto *et al.*[6] favour the latter possibility and claim that short stretches of DNA, the so-called "Okazaki fragments", are produced individually at the growing points of the parent chains and there are joined together by the enzyme *polynucleotide ligase*. This enzyme, polymerase I and possibly polymerase II and others perform an additional and important task in keeping the DNA molecule intact. Cells normally contain nucleic-acid-hydrolysing enzymes, the endo- and exo-nucleases, and due to their action breaks in the DNA chains may occur. Those breaks are repaired by the above enzymes, thus restoring the structural and biological stability of the genetic material.

CHAPTER 4

Ribonucleic Acid and Transcription

AFTER having dealt with DNA replication we now proceed to describe how DNA-contained genetic information is transcribed, i.e. how a sequence of DNA nucleotides is transferred to a complementary sequence in a molecule which may leave the nucleus and pass into the cytoplasm, there to be translated. The initial step in this process has been termed transcription.

Jacob and Monod assumed the existence of a certain type of RNA whose function is the transfer of such information to the ribosomes which comprise the protein synthesis apparatus, and suggested the functional name of "messenger" RNA (mRNA).

The structure of RNA, and not necessarily messenger, is essentially similar to that of DNA, the main difference being that the base uracil replaces thymine and the sugar component is ribose and not deoxyribose (see Fig. 1). The first chemical demonstration that a species of RNA resembled DNA in its base sequence was that of Volkin and Astrachan[24] who conducted their experiments on intestinal bacterial viruses. From their experiments it was apparent that the nucleic bases formed in RNA were present in the same ratio as the complementary primer DNA bases, with the above-mentioned difference that uracil and not thymidine was the complement of adenine. In other words, the principle of base pairing as outlined previously is also demonstrated in transcription and adenine, thymine, guanine and cytosine in the DNA are respectively transcribed in RNA as uracil, adenine, cytosine and guanine. Further proof of base complementation between DNA and RNA is the fact that under certain experimental conditions DNA–RNA "hybrids" are formed.

22

The DNA duplex is melted, for example, by heating, and subsequently cooled slowly in the presence of single-stranded RNA which is suspected of base complementarity. If the DNA and RNA strands zip up or *anneal* this is a proof that they are complementary. Volkin and Astrachan indeed obtained hybridization, a phenomenon which would not occur if base sequences in the DNA and RNA strands were dissimilar. Further evidence for the existence of this type of RNA is the observation that it does not anneal with itself.

Research on transcription followed in the wake of replication and an enzymatic mechanism resembling that of DNA replication was sought. This was found independently by three different laboratories, one working on plant cells (Huang *et al.*[11]), one on animals (Weiss[25]), and one on bacterial systems (Hurwitz *et al.*[12]). These workers proved the existence of the enzyme *RNA polymerase* (also called DNA–RNA polymerase) which controls DNA primed RNA formation. Not only is mRNA synthesis directed by this enzyme but all other types of cellular RNA as well. This enzyme as stated requires a DNA template and the four ribonucleoside triphosphates. With the formation of each link in the RNA chain two pyro-phosphate groups are liberated.

Despite the single name—RNA polymerase—this enzyme, as its end product the RNA itself, is not uniform and may vary in structure and function. Mondal *et al.*,[18] working on coconut nuclei, isolated two different RNA polymerases and several protein fractions from chromosomal acidic fractions, these factors conferring specific activities to each type of polymerase. Roeder and Rutter,[22] also working with eukaryotic cells, have demonstrated multiple forms of RNA polymerase, whereas in prokaryotic cells it appears that the enzyme is uniform. As outlined by Burgess[7] it has now been shown that this complex enzyme contains the following polypeptide chain subunits: one beta-prime subunit, β^1; one beta subunit, β; two alpha subunits, α; and one sigma subunit, σ. All of these subunits together, $\beta\beta^1\alpha_2\sigma$, are termed the complete or holoenzyme. The holoenzyme may be separated into two functional parts:

(a) a "core" enzyme $\beta\beta^1\alpha_2$ which is able to synthesize RNA but lacks the ability to initiate such synthesis specifically;

(b) a sigma factor σ which endows the ability of initiation RNA synthesis at specific sites.

The transcription factors

These factors, the number of which is at present steadily increasing, have marked and specific effects on the process of transcription and are associated with the core enzyme. However, they do not bind tightly enough to the core to purify with it. These factors include the psi (ψ), rho (ρ), M and CAP factors which have been determined primarily in bacterial systems but which may possibly be related to the protein factors Mondal *et al.* isolated in association with coconut polymerase.

The ψ factor is a specific factor for initiation of ribosomal RNA (of which more shall be said in Chapter 6), and increases the capacity for synthesis of ribosomal RNA by the holoenzyme several hundredfold. The ρ factor participates in the termination of synthesis of some RNA chains.

The M-factor resembles the σ-factor and is connected with initiation. Finally, mention is made of the CAP—*catabolite gene activating protein*—which is synonymous with *cyclic AMP receptor protein* or *CR protein*. The subject of cyclic AMP (cAMP) and catabolite repression will be discussed on p. 29 and at present we note in passing that this factor is involved in mediating the effect of cAMP which is required for the expression of a certain class of genes.

It is of interest to note that as opposed to the process of DNA replication, in transcription only one chain of the duplex is active (see also Fig. 25 on p. 66). The two strands therefore possess a certain asymmetry the nature of which is still unknown. This phenomenon was furthermore shown in pea tissue by Bonner *et al.*[6] who found that during the initial stages of RNA polymerase action a nucleic complex was obtained in which the ratio of DNA to RNA was 2:1. Apart from this in a hybridization experiment performed as described previously, with radioactive RNA it was found that only one of the duplex DNA strands was able to hybridize with the labelled RNA.

Another question that presents itself in transcription is whether the primer for RNA synthesis is DNA in its unwound state or whether unwinding of the segment to be transcribed is a prerequisite for the process. It has been commonly believed that DNA unwinds during replication and more locally during transcription. In the latter process it has been suggested that local melting of DNA occurs on the binding

PLATE 2

Visualization of transcription. The historic photograph of Miller and Beatty.[17] This electromicrograph ($\times 25,000$) of the oocyte of the gnat *Triturus viridescens* indicates long axes which are DNA strands. Numerous shorter strands being RNA chains branch from the DNA axes. It is noted that not all of the DNA is active and there are regions lacking RNA. A general calculation indicated that only about 10 per cent of the total observed DNA produced RNA, this being in accordance with current concepts of repression and derepression. At the points of junction of the RNA strands to the DNA axes little spheres can be observed, these are believed to be molecules of the enzyme RNA polymerase. (Reproduced with the kind permission of O. Miller and B. Beatty, Dept. of Biology, Oak Ridge National Laboratories, Tennessee, U.S.A.)

of the RNA polymerase to DNA and then this open unwound region moves along the transcribed strand with the enzyme. Claims have even been made for the existence of an "unwindase" enzyme although because of the experimental difficulty of demonstrating the local unwinding in transcription the possibility that the process may occur on intact double helical DNA has not been ruled out.

The units of mRNA synthesized on the DNA primer are considerably smaller than the primer DNA molecule. According to its informational content the mRNA chain varies in length (see Plate 2) and is therefore, as an RNA species, heterogeneous in molecular weight which within very broad boundaries ranges between the high m.w. of ribosomal RNA-0·5–5 millions and the low m.w. of transfer RNA which possesses a m.w. of approximately 25,000. Of these latter two RNA species more shall be said presently. The longevity of mRNA molecules also varies with species and is possibly related to the degree of species organization and evolution. The average half-life of bacterial mRNA is very short indeed—only a few minutes—while mRNA of higher organisms is relatively stable. RNA may be synthesized at several sites at once on various sections of the DNA chain and the polymerase continues to traverse along the strand assembling nucleotides according to the DNA-directed base sequence. The enzyme ceases to transcribe and is set free from the chain when a termination signal is encountered. This signal has been linked with the protein ρ (rho) factor in conjunction with the RNA polymerase itself and may depend on specific base sequences at the termination sites such as the ochre and amber codons (see p. 39) on the primer.

Another enzyme capable of RNA synthesis is *polynucleotide phosphorylase* discovered by Grunberg-Manago and Ochoa in 1955. During the process of its action inorganic phosphate is released and RNA with a completely random base sequence is formed. DNA as a primer is not required and if supplied with a random pool of nucleoside diphosphates the RNA synthesized contains no genetic information. Great use has, however, been made of RNA synthesized with this enzyme when supplied with a single nucleoside diphosphate and this has facilitated the breakdown of the Genetic Code which will be discussed in the next chapter.

TRANSCRIPTIONAL CONTROL MECHANISMS

A. *The operon*

The present operon concept of transcriptional control in protein synthesis is with certain modifications, as hypothesized by the French biologists Jacob and Monod and in general follows the outlines summarized by them in 1961.[14] When this concept was originally suggested certain stages were at that time purely conjectural but have since been vindicated. In bacterial systems the operon has been shown to be one of the major control mechanisms, but in higher plants and animals the situation may be more complex.

With the knowledge that protein—the end product of informational transfer from the DNA—is manufactured in the cytoplasm while the genetic information is nucleus-contained, Jacob and Monod assumed the intervention of a connective link, formed on the genome and transferable to cytoplasm and suggested that messenger RNA, discussed above, performs this task. Analysis of the kinetics of enzyme (protein) induction and repression indicated that an enzyme's production could be induced by its substrate or by a specific metabolite, and in this event the process is termed *positive enzymatic adaptation*. On the

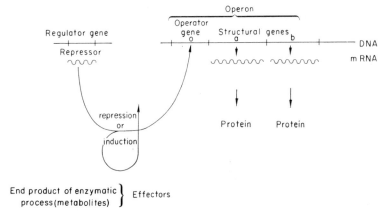

Fig. 11. The original Jacob and Monod operon model: *a* and *b* are structural genes; O is the operator site of repressor attachment.

other hand, it was also noted that enzyme production could be *repressed* by its product or another specific metabolite and in this case *negative enzymatic adaptation* is encountered, and upon removing the metabolite enzyme synthesis is resumed.

Using these observations, Jacob and Monod suggested that in the apparatus of informational transfer appropriate to one final protein product, there exists a separate gene region which regulates the informational transfer, and this separate gene they designated the *regulator gene*. This gene according to the scheme shown in Fig. 11 produces a *repressor* which is able to inhibit the transcription of information contained in a specific segment of the chromosome. The repressor may be affected by end-products of enzymatic action. If a metabolite accentuates the repressor's effect the phenomenon is termed *co-repression*, but if the metabolite promotes enzyme synthesis and overcomes inhibition the effect is one of *derepression*. The regulator gene may control the synthesis of a single protein or may control synthesis of several proteins.

Jacob and Monod consequently deduced that the repression hypothesis leads to the assumption that an operator gene exists and the problem of its site and mode of operation was posed. The site is believed to be at a specific position on a chromosomal segment called the *operon* which in addition to the operator gene also contains several *structural genes* sequentially appearing on the strand and which induce the formation of a series of enzymes all of which participate in a given metabolic pathway; for example, the fermentation of lactose in *E. coli* the inducer being the substrate itself—lactose—and hence the name the *lac operon*.

The three enzymes participating in lactose fermentation are in order of their action: β-galactosidase, β-galactoside-permease and β-galactoside-transacetylase and the three structural genes coding for these proteins are next to each other in one operon on the *E. coli* chromosome.

In the micro-organism *Salmonella*, for example, histidine synthesis is dependent upon nine different enzymes, the genes for the production of which are found together or "clustered" sequentially along a given chromosomal segment. A mutation at the beginning of the segment may have a general or pleiotropic effect on the ability to synthesize all nine of the enzymes. This is understood to be an indication that this series of genes comprises an operon controlled by an operator adjacent to a certain cistron. It is, however, possible that the structural genes coding for a given metabolic pathway are not clustered and may be present on diverse sites along the chromosome or even on different chromosomes.

Subsequent work modifying the original operon concept has focused attention on the operator region which according to the above scheme is the site of attachment of the repressor. Research in this context has also included the relationship of the operator and the attachment of RNA polymerase responsible for subsequent mRNA synthesis. Ippen *et al.*[13] have described a further site on the operon designated *promotor* which is believed to be the attachment site of RNA polymerase on the chromosome. According to this theory and as outlined in Fig. 12, the

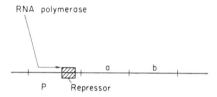

Fig. 12. A modification of the operon model as described by Ippen:[13] P = promotor and is a specific locus on the operator gene and is the site of attachment of the RNA polymerase.

enzyme is attached to the left of the site of attachment of the repressor and the latter, if present, serves to prevent the progress of the enzyme along the chromosome, thus inhibiting transcription.

With the discovery of cyclic adenosine monophosphate (cAMP) whose role in hormonal mechanisms will be discussed in detail in Part Two of this book and the isolation of the RNA polymerase-associated CAP factor mentioned on p. 24, the operon concept has been further modified. For several decades the "glucose effect" on enzymes has been known. This is the observed ability of glucose to inhibit the synthesis of many but certainly not all enzymes. The term "catabolite repression" has now been adopted in its stead, since the effect is not limited to glucose but also includes its degradation products or catabolites as well as glucose-related compounds like gluconic acid, mannitol and galactose (Magasanik[16]). More recently Pastan and Perlman[20] have shown that cAMP stimulates enzyme synthesis in cultures repressed by glucose. Although cAMP stimulated synthesis of a widely diverse group of enzymes including those participating in carbohydrate transport and metabolism, amino-acid and pyrimidine metabolism its action was still specific. The nucleotide had no effect on enzymes not subject to catabolite repression and furthermore had no effect on total RNA synthesis. Additionally, it has been demonstrated that cyclic AMP controls the synthesis of all (substrate) inducible enzymes and in general prevents catabolite repression of glucose-sensitive enzymes.

On the grounds of these and other observations, Pastan and Perlman presume that cAMP directs RNA polymerase to initiate transcription at the promotor. This is achieved indirectly by means of combination of cAMP with an accepting protein—the "catabolite-gene activator protein" or CAP factor mentioned previously. The complex cAMP-CAP seems to act at the promotor site since mutants lacking the site do not respond to it. However, it is as yet not clear whether the complex produces some change in the RNA polymerase (arrow 1 in Fig. 13) or in

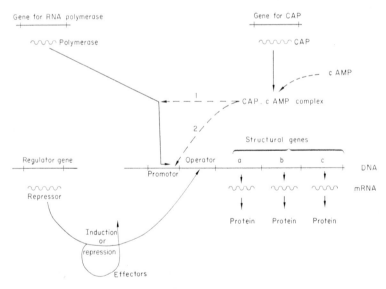

Fig. 13. Pastan and Perlman's[20] reconstruction of the operon model in cyclic AMP-sensitive metabolic pathways. See text for details.

the DNA (arrow 2). In either event the result is an increased synthesis of mRNA due to increased frequency of initiation of mRNA chains. Finally it should be mentioned that CAP protein resembles the σ-factors associated with the polymerase and which also stimulate the initiation of transcription.

The nature of the molecular repressor. When Jacob and Monod originally suggested their model, the existence of repressors was purely hypothetical, even though they did answer well for many observed phenomena. It was even suggested that they might be nucleic acid entities. Only in 1966 with the lac system did Gilbert and Müller-Hill[10] succeed in chemically isolating such an entity. This was achieved by utilizing the metabolite isopropyl thiogalactoside (IPTG). This substance causes induction of synthesis of the first enzyme of the metabolic sequence involved in lactose fermentation—β-galactosidase. The above workers determined that the repressor in this system is a protein having a molecular weight of 150–200,000 and is comprised of several subunits. This repressor and another isolated from the λ-phage by Ptashne[21] possess an affinity for the operator on the DNA. The exact nature of the affinity is not known but amongst other factors electrostatic attraction plays an important role. It is assumed that the attachment of the repressor to DNA is reversible, i.e. after attachment to DNA under certain conditions it may be freed thus allowing mRNA synthesis.

The above pertains to the binding of repressors to DNA and as regards the binding to the effector or metabolite (see Fig. 12) Monod *et al.*[19] suggested that this may depend on the steric structure of the protein repressor. They claim that the process is stereospecific, rapid and reversible. According to their model only one gene is responsible for the production of the repressor which is a macromolecule of considerable dimensions, this being in accordance with the experimental findings reported above. It has, moreover, been shown that a single mutation which prevents the repressor from reacting with the effector does not prevent its binding to DNA. From these data they suggested that repressors are comprised of an "allosteric" protein. This concept requires a protein with two active sites, one of which combines with the operator region of DNA and the other with the effector. The allosteric mechanism of repressor control is similar to that of the allosteric regulation of activity of certain enzymes and of oxygen uptake by blood haemoglobin (Changeux[8]). Figure 14 shows the mode of allosteric control in general outline. It is assumed that the linkage of repressor to effector is not covalent but rather of weaker forces, such as van der Waals. The nature of the linkage may account for the comparatively speedy reversibility of the above process.

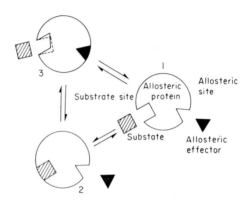

Fig. 14. Allosteric regulation of biologically active proteins. The allosteric effector (or metabolite) upon combination with the protein causes a steric change of the substrate site thus preventing protein–substrate reaction.

As seen in Fig. 14 it is evident that the allosteric effector has no connection with the substrate and that it binds directly with the protein.

Growth hormones and inhibitors may in certain cases act as allosteric effectors. It is possible that these act on enzymes equipped with allosteric sites and called *regulatory enzymes,* or possibly action is at the level of the gene repressor. This mode of hormone action will be discussed in Part II.

Under certain conditions the sensitivity of a regulatory protein to allosteric effectors may be abolished without impairing the ability to act on the substrate. This may be achieved by heating and subsequent cooling under controlled conditions, or by application of mercury salts, e.g. para chloromercurobenzoate (Stadtman[23]). This fact may offer a physiological explanation for the phenomenon of "hardening" and adaptation of plants and other organisms to extreme conditions of temperature or chemical environment.

B. *The histone control mechanism*

Whereas the operon hypothesis still awaits demonstration in higher organisms, the histone transcriptional control mechanism has a firmer basis and is more widely recognized although not unanimously accepted. According to Bonner[3] it is possible that in higher plants one of the main factors preventing DNA transcription is a certain type of protein known as histone. In several differentiation processes, e.g. flower differentiation from vegetative primordia in *Chenopodium alba*, there is histochemical evidence that the content of histone decreases in the differentiating buds and that at the same time RNA content increases. Furthermore, in ryegrass, *Lolium temulentum*, it was shown that the inhibitor actinomycin D, which prevents flower differentiation under normally inductive long-day photoperiods, combines with histone-free DNA. It is believed that during the process of flower differentiation derepression of flowering genes is achieved by histone removal.

Contrary to the above findings reports have been made on other plants where flower-bud differentiation was not accompanied by histone decrease. In these cases, however, it was found that qualitative histone changes occur and a reduction in arginine-rich histone was found. The arginine-rich histone is believed to be more inhibitory than lysine-rich histone and experiments performed on pea plants showed that auxin combines especially with arginine-rich histones, this suggesting a derepressive action of the hormone.

In view of the above it appears that histone is involved in the control of hormone-mediated growth. Histones in general are bound to DNA and the ratio of histone to DNA is specific for various plant and animal species. It is also believed that histones may determine the degree of coiling of the double helix.

The activity of enzyme RNA polymerase has also been shown to be inhibited by histone. Allfrey *et al.*[1] have suggested that it is possible to reverse histone-induced genetic repression by local *acetylation*—the agents performing this task being the histone acetylases. These workers found a complete reversal of RNA polymerase inhibition caused by histones when the latter were acetylated. Furthermore, the "puffs" in the giant chromosomes of insect salivary glands believed to be transcribing derepressed DNA (see p. 100 and Fig. 34) contain histone acetylase. Allfrey *et al.*[2] conclude that this enzyme is an important link in DNA derepression and may be activated by growth hormones like ecdysone, the insect moulting hormone which also causes puffing.

Concerning the direct participation of histone in DNA repression phenomena Bonner[4] has posed an interesting question. It is known that the linkage of histone to DNA is a simple ionic bond and no specificity exists between histone types and DNA. This implies that histone does not "seek out" DNA or specific DNA segments and the combination does not depend upon the informational content of the DNA. The histone is unable to read the DNA base sequence and the mode of associating of the two molecules is the affinity of positively charged histone to the negatively charged DNA. The obvious question is therefore how is the histone able to function as a selective repressor and how can localized derepression be reconciled with the lack of specificity of the histone/DNA bond. The answer to this problem is the discovery by Bonner's team[5] of *chromosomal RNA*. This is a type of RNA clearly distinct from either mRNA, ribosomal RNA or transfer RNA. It is hydrogen bonded to the chromosome and at the same time covalently bonded to "initial peak protein" which in turn is bonded to several histone molecules. This chromosomal RNA is a comparatively small molecule comprised of only forty to sixty nucleotides having a coefficient of sedimentation of 3·2 S. This RNA according to Bonner may be the means whereby the histone recognizes specific base sequences on the DNA.

Finally, it is mentioned that inasmuch as the histone hypothesis is valid, it is only so for transcription of nuclear DNA in higher plants and animals. Bacterial DNA and that of eukaryotic organelles such as mitochondria and chloroplants is considered to be "naked" and is not complexed with histone.

RECOMMENDED READING AND REFERENCES

1. ALLFREY, V. G., FAULKNER, R. and MIRSKY, A. E., Acetylation and methylation of histones and their possible role in the regulation of RNA synthesis. *Proc. Nat. Acad. Sci.*, **51**, 786–94 (1964). See no. 2 following.

2. ALLFREY, V. G., POGO, B., LITTAN, V., GERSHEY, E. and MIRSKY, A. E., Histone acetylation in insect chromosomes. *Science*, **159**, 314–16 (1968). A research report which together with no. 1 above indicated the importance of enzyme participation in "puffing", thus suggesting that hormone action possibly is not direct but enzyme mediated.

3. BONNER, J. (1965). See no. 2, Chap. 2.

4. BONNER, J., The next new biology. *Proc. 17th Int. Hort. Cong.*, **2**, 45–59 (1966). A brief and pertinent summary of transcriptional control mechanisms and future research horizons.

5. BONNER, J., DAHMUS, M., FAMBROUGH, D., HUANG, R., MARUSHIGE, K. and TUAN, D. The biology of isolated chromatin. *Science*, **159**, 47–56 (1968). An excellent description, stressing methodology, of chromatin action and isolation. Includes an interesting discussion on the histone hypothesis and the newly discovered chromosomal RNA.

6. BONNER, J., HUANG, R. C. and MAHESHWARI, N. The physical state of newly synthesized RNA. *Proc. Nat. Acad. Sci.*, **47**, 1548–54 (1961). Proof, from plant systems, that during transcription only one DNA strand is utilized.

7. BURGESS, R., RNA polymerase. *Ann. Rev. Biochem.*, pp. 711–40 (1971). An up-to-date review of this key enzyme and all its factors.

8. CHANGEUX, J. P., The control of biochemical reactions. *Scient. Am.*, **212**, 36–45 (1965). A semi-popular article demonstrating, with the aid of excellent illustrations, several possible modes of control of cellular activity.

9. EVANS, L. T., Inflorescence induction in *Lolium temulentum* VI. Effects of some inhibitors of nucleic acid protein and steroid biosynthesis. *Austral. J. Biol. Sci.*, **17**, 24–35 (1964). A series of experiments investigating the possible link between nucleic acid metabolism and flower differentiation. See also no. 15 below.

10. GILBERT, W. and MÜLLER-HILL, B., Isolation of the lac repressor. *Proc. Nat. Acad. Sci.*, **56**, 1891–8 (1966). The first experimental proof of this repressor's existence.

11. HUANG, R. C., MAHESHWARI, N. and BONNER, J., Enzymatic synthesis of RNA. *Biochem. Biophys. Res. Comm.*, **3**, 689–94 (1960). Characterization of RNA polymerase in plant systems.

12. HURWITZ, J., BRESLER, A. and DIRINGER, R., The enzymic incorporation of ribonucleotides into polyribonucleotides and the effect of DNA. *Biochem. Biophys. Res. Comm.*, **3**, 15–19 (1960). Characterization of RNA polymerase in bacteria.

13. IPPEN, K., MILLER, J. H., SCAIFE, J. and BECKWITH, J. R., New controlling element on the Lac operon of *E. coli*. *Nature*, **217**, 825–7 (1968). An important modification of the "operon" concept.

14. JACOB, F. and MONOD, J., Genetic regulatory mechanisms in the synthesis of proteins. *J. Molec. Biol.*, **3**, 318–56 (1961). A now classic document on the control of protein synthesis. A detailed explanation of the "operon" model.

15. KNOX, R. B. and EVANS, L. T., Inflorescence initiation in *Lolium temulentum* L. VIII. Histochemical changes at the shoot apex during induction. *Austral. J. Biol. Sci.*, **19**, 233–45 (1966). See no. 9 above.

16. MAGASANIK, B., Catabolite repression. *Cold Spring Harb. Symp. Quant. Biol.*, **xxvi**, 249–56 (1961). A logical explanation of the process stressing the rationale of the suggested new terms.

17. MILLER, O. L. and BEATTY, B. R., Visualization of nuclear genes. *Science*, **164**,

three or six nucleotides are added or removed a functional end-product is obtained. Results of experiments conducted along these lines indicated that codons are indeed triplets and that translation proceeds from a given point onwards.

Nevertheless, we still may ask how the sixty-four possible triplet combinations (codons) of the four basic nucleotides produce only twenty amino acids. The answer is that when the genetic code (i.e. the determination of which sequence leads to which amino acid) was discovered it became apparent that certain amino acids could be coded for by more than one triplet. This phenomenon is termed *degeneracy* or *redundance*.

The genetic code

The unravelling of the genetic code was initiated by Nirenberg and Matthaei[5] who, with the aid of the enzyme polynucleotide phosphorylase mentioned in the previous chapter, synthesized polynucleotide chains. By varying the nucleotide composition of these chains and using them as synthetic mRNAs several codons were discovered. For example, the polyribonucleotide obtained by polymerizing uracil-poly U, when used as mRNA, always coded for the amino acid phenylalanine or, to be more exact, a polypeptide chain consisting of polyphenylalanine. The conclusion drawn from this experiment is that the codon or one of the codons for phenylalanine is UUU and that a homopolynucleotide codes for a homopolypeptide. This approach, subsequently extended by the synthesis in Khozana's laboratory of mixed ribo-oligonucleotides of defined sequence, enabled codons for most amino acids to be determined. Table 2 summarizes the genetic code as known at present. The code is believed to be "universal", i.e. the same codons spell the same amino acids in all living organisms, thus indicating a certain Unity of Creation.

Table 2 indicates that the codons for phenylalanine may be UUU or UUC. A similar situation exists for several other amino acids which may be coded for by more than one codon. However, further scrutiny of these instances reveals that the first two bases in each such case of degeneracy are identical and change is obtained only by variation of the third base in the triplet. According to Crick[2] in these cases a certain *wobble* exists in the choice of the third base and this wobble may be

TABLE 2. THE GENETIC CODE AS CONTAINED IN mRNA

Ochre and amber are codons for chain termination, and the codons AUG and GUG may act in the process of initiation. * UGA is the codon for cysteine in vertebrates, while in bacteria it is a termination codon.

Nucleic base at second position in codon

		U	C	A	G	
		phenylalanine	serine	tyrosine	cysteine	U
	U	phenylalanine	serine	tyrosine	cysteine	C
		leucine	serine	ochre	cysteine*	A
		leucine	serine	amber	tryptophan	G
		leucine	proline	histidine	arginine	U
	C	leucine	proline	histidine	arginine	C
		leucine	proline	glutamine	arginine	A
		leucine	proline	glutamine	arginine	G
		isoleucine	threonine	asparagine	serine	U
	A	isoleucine	threonine	asparagine	serine	C
		isoleucine	threonine	lysine	arginine	A
		methionine	threonine	lysine	arginine	G
		valine	alanine	aspartic acid	glycine	U
	G	valine	alanine	aspartic acid	glycine	C
		valine	alanine	glutamic acid	glycine	A
		valine	alanine	glutamic acid	glycine	G

Nucleic base at first position in codon · *Nucleic base at third position in codon*

accounted for by the nature of hydrogen bonding between two complementary bases. This bonding as seen in Fig. 5 on p. 13 may either be of two links or of three, the former possibility providing the "wobble". He therefore considers that a perfect fit between the bases in the codon and the anti-codon (to be discussed presently) is not an essential pre-requisite for mutual recognition. The anticodon region in tRNA often contains one of the less-frequent nucleic bases, inosine, and this base when in the third position may "recognize" three of the regular bases, U, C or A, in the third position of the codons.

Transfer RNA. The genetic code may be regarded as a dictionary used by the cell to translate a four-lettered language of nucleotides into a twenty-lettered language of amino acids. The interpreters of the process are specialized "adaptor" molecules whose RNA functions in trans-ferring amino acids—hence the name *transfer* RNA or tRNA. The term soluble RNA (sRNA) often used is inclusive of tRNA but may contain other RNA types as well—solubility relating to the characteristic property of this fraction in not precipitating in 1 N NaCl solutions. This

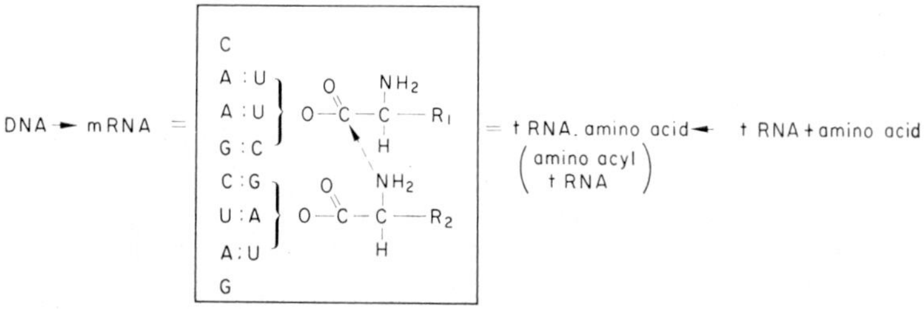

Fig. 15. The codon–anti-codon or "adaptor" hypothesis. It can be seen how two amino-acid-loaded tRNA molecules are arranged along the mRNA. By complementation of the nucleic bases the tRNA anti-codons are fitted on to the mRNA codons. The arrow indicates how two amino acids subsequently combine. (After Zachau[7].)

type of RNA is also referred to as "light" RNA since its molecular weight approximates 25,000, being only of about 80 nucleotides in length and having a sedimentation coefficient in Svedberg units of 4 S.

Each amino acid has a separate tRNA molecule containing a specific nucleotide sequence complementary to that in the codon on the mRNA and also specific for the particular amino acid carried by the tRNA molecule (see Fig. 15). The complementary region in the tRNA molecule is termed the *anti-codon*. Like mRNA, tRNA too is assembled by RNA polymerase acting on the DNA primer, but a certain degree of alteration of the bases may occur. These may be thiolated (SH groups added), methylated (CH_3 groups), formylated (CHO added), etc., and as mentioned previously some of the major bases may be replaced by less frequent or minor ones such as inosine, pseudouridine and others. When the tRNA molecule is arranged in such a way as to maximize the hydrogen-bonding between its constituent bases, the structure resembles a clover leaf and this term has been coined by molecular biologists to describe tRNA configuration. Figure 16 shows two types of tRNA whose base sequences have been determined. It is also assumed that the double-stranded regions have a double helical configuration. Base sequences of several tRNA's have been determined and it is now possible to outline general structural principles of tRNA's. According to Schreiber[6] all tRNA's are clover-leafed in structure and have five arms:

1. the amino-acid arm or stem;
2. the dihydrouridine arm;
3. the anti-codon arm;
4. the thymine-pseudouridine arm;
5. the "extra" arm, which is of variable lengths.

These arms can be seen in Fig. 16. It can also be seen that the molecule terminates with a CCA sequence. This is a common factor to all types of tRNA molecules and the enzyme responsible for its addition to the tail end of *all* the various types of tRNA molecules is *tRNA pyrophosphorylase*. The CCA tail is essential for the binding of amino acids to the molecule and this is done by linking a hydroxy group (OH) of the terminal adenylate of the tail to the carboxy (CO–) terminal of the amino acid specific to the particular kind of tRNA being "loaded". As stated before the CCA terminal is universal for tRNA molecules but the sequence of bases proximal to this tail is believed to be specific for the different types of amino acids and may, moreover, vary between different organisms for the same amino acid.

One of the substituted bases, isopentenyl adenine, which often appears in tRNA also belongs to the cytokinin group of plant hormones. The possibility that hormonal action may be attributed to the presence of the isopentenyl adjacent to the anti-codon in tRNA will be discussed in the chapter dealing with cytokinins in Part Two of this book.

The enzymes participating in the binding of amino acids to tRNA are

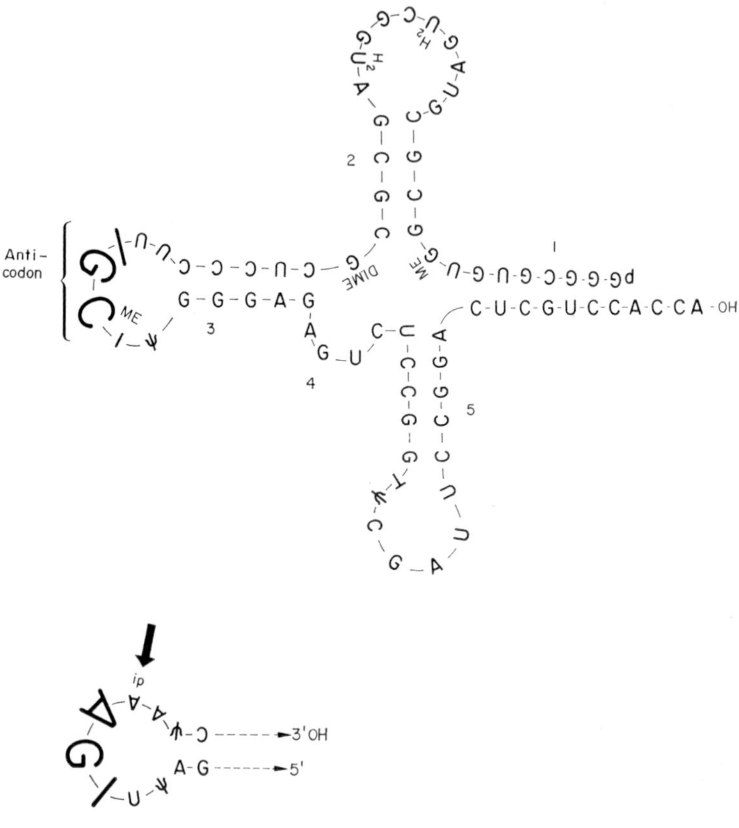

FIG. 16. *Above:* The proposed clover-leaf arrangement and base sequence of alanine tRNA. (After Holley *et al.*) Me = methyl, ψ = pseudouridine, I = inosine. The five arms according to Schreiber[6] are: 1. the amino-acid arm, 2. the dehydrouridine arm, 3. the anti-codon arm, 4. the "extra" arm, 5. the thymine pseudouridine arm. *Below:* The anti-codon loop of serine tRNA. Adjacent to the anti-codon the base isopentenyl adenine (ipA—indicated by arrow) is found. This base has cytokinin activity.

called *amino-acyl tRNA synthetases* and it is believed that each amino acid has its specific synthetase. The steps of loading tRNA with an amino acid are as follows:

1. Amino acid + ATP + synthetase
   $\leftrightharpoons$ amino acid·AMP·synthetase + PP (inorganic pyrophosphate)
   (complex)
2. Complex + tRNA $\leftrightharpoons$ amino acid·tRNA + AMP + synthetase

According to Doctor *et al.*[3] the binding of the amino-acyl synthetase to the tRNA may occur in one of the following ways:

(a) The enzyme may recognize a certain base sequence somewhere along the tRNA molecule (possibly the sequence adjacent to the CCA "tail"—author).

(b) A steric fit has to be found between certain regions of the enzyme and the tRNA.

(c) A steric fit between the complete enzyme molecule and the specific tRNA has to be found.

Doctor *et al.* propose that the steric compatibility may be of more importance than base complementation in this case. The problem of recognition between specific synthetases and each of the twenty different amino acids has been termed "The Second Genetic Code" and at present is one of the central mysteries of modern molecular biology.

RECOMMENDED READING AND REFERENCES

1. CRICK, F. H. C., The genetic code. *Scient. Am.* (Oct. 1962). This article together with nos. 2 and 4 below explain how, using different approaches, various groups of scientists unravelled the genetic code.
2. CRICK, F. H. C., The genetic code III. *Scient. Am.* **215,** 55–62 (1966).
3. DOCTOR, B., LOEBEL, T., SODD, M. and WINTER, D., Nucleotide sequence of *Escherichia coli* tyrosine transfer ribonucleic acid. *Science,* **163,** 692–5 (1969). The possibility of existence of a second genetic code is suggested, and mechanism of tRNA binding to amino acid discussed.
4. NIRENBERG, M. W., The genetic code II. *Scient. Am.* (Mar. 1963).
5. NIRENBERG, M. W. and MATTHAEI, J. H., The dependence of cell free protein synthesis in *E. coli* upon naturally occurring or synthetic polyribonucleotides. *Proc. Nat. Acad. Sci.,* **47,** 1588–1602 (1961). A major breakthrough towards the understanding of the genetic code.
6. SCHREIBER, G., Translation of genetic information on the ribosome. *Angew. Chem., Int. English Ed.,* **10,** 638–51 (1971). A thorough, well-illustrated review of the process of translation.
7. ZACHAU, H. E., Transfer amino acids. *Angew. Chem., Int. English Ed.,* **8,** 711–27 (1969). tRNA configurations, base sequences and methods of determination described in detail.

CHAPTER 6

Ribosomes and Polymerization of tRNA-borne Amino Acids

THE process of polymerization which finally results in the appearance of a polypeptide chain occurs on the ribosomes which can be likened to protein factories. The role of ribosomes aided by certain enzymes is analogous to that of RNA polymerase and DNA polymerase. All three are means whereby "nucleic" acid template-contained information is interpreted.

A typical ribosome is built of approximately equal parts of RNA and protein. Size is often expressed not as molecular weight but as Svedberg (S) units, i.e. sedimentation coefficients calculated from speed of sedimentation upon ultracentrifugation. Higher plants and animals have 80 S ribosomes whereas lower organisms such as bacteria have 70 S ribosomes. Figure 17 indicates the division of the large unit into smaller subunits.

Consider eukaryotic ribosomal RNA: it is now believed that the precursor ribosomal RNA is 45 S, has a m.w. of $\sim$ 4·3 million, and is formed on DNA template in the region of the nucleolus. This larger unit is enzymatically split into 25 S (or 28 S) and 18 S subunits and to these, while still in the nucleus, the protein is added, thereby increasing their bulk to 60 S and 40 S consecutively. These subunits are then discharged into the cytoplasm together with smaller 7 S and 5 S units. In plants it has been reported that different families may have a ribosomal precursor (usually 45 S) of a somewhat different molecular weight and interestingly enough this may also vary within one and the same plant. For example, ribosomal precursors in roots and leaves may differ considerably but it appears that the final units which ultimately comprise the complete plant ribosome are the same. In bacteria such as *E. coli* the

44

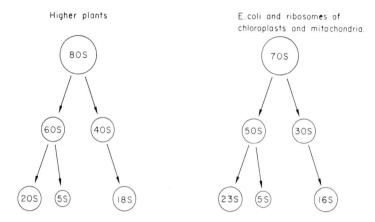

Fig. 17. The division of the ribosome into its subunits.

subunits are somewhat smaller and the large subunits are 50 S and 30 S.

Ribosomes in cells may appear singly and are then termed mono-somes, or may be "threaded" as a group on strands of mRNA and are then called "polysomes" (see Plate 3, p. 47). In some organisms the poly-somes are associated with the endoplasmic reticulum membrane while in others, e.g. tobacco pith or corn roots, they exist separately.

By polyacrylamide gel electrophoresis, Loening[5] has compared molecular weights and S values of RNA from the two basic subunits of various prokaryotic and eukaryotic organisms. Table 3 summarizes his results.

These and other results show that bacteria, Actinomyces, blue-green algae and higher plant chloroplasts all have rRNA (23 S and 16 S) with molecular weights of 1·1 million and 0·56 million. The weights of the rRNA (25 S and 18 S) of the higher plants, ferns, algae and fungi and of some protozoa are approximately 1·3 million and 0·7 million. The 0·7-million component is also common to all animals, but the larger rRNA component (28 S) has evolved with each major step of animal evolution from about 1·4 million in sea urchins to 1·75 million in mam-mals. Loening attributes evolutionary importance to these findings and it indeed seems that evidence is presented in support of the hypothesis that the chloroplast is a prokaryotic "invader" into an eukaryotic host.

TABLE 3. COMPARISON OF MOLECULAR WEIGHTS AND S VALUES OF RNA TYPES IN RIBOSOMAL RNA OF VARIOUS ORGANISMS

(According to Loening)[5]

Ribosome source	m.w. $\times 10^6$	
	In the larger subunit	In the smaller subunit
Bacteria	1·10 (23 S)	0·56 (16 S)
Actinomyces	1·10 (23 S)	0·56 (16 S)
Blue-green algae	1·10 (23 S)	0·56 (16 S)
Chloroplasts of higher plants	1·10 (23 S)	0·56 (16 S)
Ferns	1·30 (25 S)	0·70 (18 S)
Protozoa	1·30 (25 S)	0·70 (18 S)
Higher plants	1·30 (25 S)	0·70 (18 S)
Sea urchins	1·40 (28 S)	0·70 (18 S)
Mammals	1·75	0·70 (18 S)

The chloroplast, as will be described in Chapter 7, contains DNA but has no definite nucleus or nuclear region. Loening also proposes that increase of size of the larger component may be connected with greater differentiation capacity of the higher organisms accompanied by a loss of totipotency of the differentiated tissues, the totipotency being retained in lower forms of life.

The steps of polypeptide chain formation are as follows:

(a) initiation,
(b) elongation,
(c) termination.

Initiation

Chemical analyses of cellular proteins have indicated that 95 per cent of the *N*-terminal amino groups are methionine, alanine, serine or threonine, i.e. all are somehow connected with methionine metabolism. This observation led to the assumption that a methionine derivative may assume a central role in the process of initiation. Marcker and Sanger[6] and others have shown that polypeptide chain initiation commences by translation of two codons, GUG and AUG, which code for a specific type of tRNA—the *N*-formyl-methionine

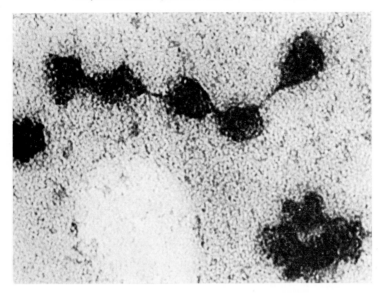

A polysomal unit in a rabbit reticulocyte cell. It can be seen that the individual
ribosomes are interconnected with a strand which is mRNA. (Reproduced
by the courtesy of Prof. A. Rich, M.I.T., U.S.A.)

tRNA. This tRNA which carries the amino-acid methionine in a
formylated (formic acid added) state is found in "cell-free" protein
synthesizing extracts, and all proteins produced by such extracts contained such formylated methione groups at the *N*-terminal. This type of
tRNA was found in *E. coli* extracts and it was shown that the formyla-
tion occurs after methionine has been esterified by tRNA.

If the above is true, one may ask why do not all proteins have *N*-
formyl-methionine or at least methionine as their *N*-terminal amino
acids? Actually, initially and in cell-free extracts they do, but *in vivo*
formate, formyl-methionine or even more amino-acid groups may
subsequently be removed. It is considered that since *N*-formyl-methio-
nine tRNA cannot form a peptide bond through its amino group
blocked by formate, it conceivably is able to serve both as a terminator
and an initiator.

Apart from *E. coli*, the organism with which the initial research was conducted, this *N*-formyl-methionine tRNA has been found in other bacteria, and also in mitochondria and chloroplasts of higher plants, but *not* in the latter's cytoplasm. This lends further support to the previously stated hypothesis that the higher plant is a symbiont comprised of eukaryotic and prokaryotic components.

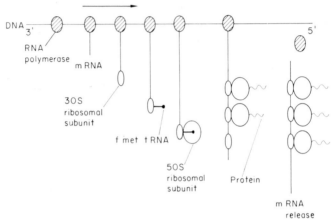

FIG. 18. The formation of the polysome in *E. coli*. (After Nirenberg[10].)

The process of polysome assembly is outlined in Fig. 18. It can be seen that initially while the mRNA still clings to the DNA, the smaller (30 S) ribosomal subunit attaches, thereafter followed by formyl methionine tRNA and only then does the larger (50 S) subunit join the complex. The complete ribosome moves up the mRNA chain allowing attachment of another unit and finally the completed polysome and the RNA polymerase are released from the DNA and protein synthesis continues. The attachment of the ribosomal subunits is aided by specific protein factors—the *initiation* factors (Revel *et al.*[12]) which have been isolated from ribosomes by extraction in 1 M NH_4Cl. In bacteria three such factors, F_1, F_2 and F_3, have been isolated. F_1 and F_2 aid binding of the tRNA to the ribosome, the latter factor also having GTPase activity. F_3 aids the attachment of the ribosome to mRNA. That such factors are not only present in bacterial systems but exist in higher plants as well has been shown by Marcus[7] using a ribosome-supernatant source from wheat embryos and tobacco mosaic virus RNA as messenger. It should be mentioned that hormone action in certain instances, e.g. of oestrogens, has been partially attributed to effects, possibly allosteric, on initiation factors.

Elongation

Figure 19 summarizes the process of elongation. It can be seen that there are two sites of activity of the tRNA participating in amino-acid polymerization. Site "A", or the "acceptor site" to which the tRNA carrying amino acid attaches, and site P, the "donor" or "peptidyl" site at which the tRNA gets rid of its load. Peptide chain growth commences

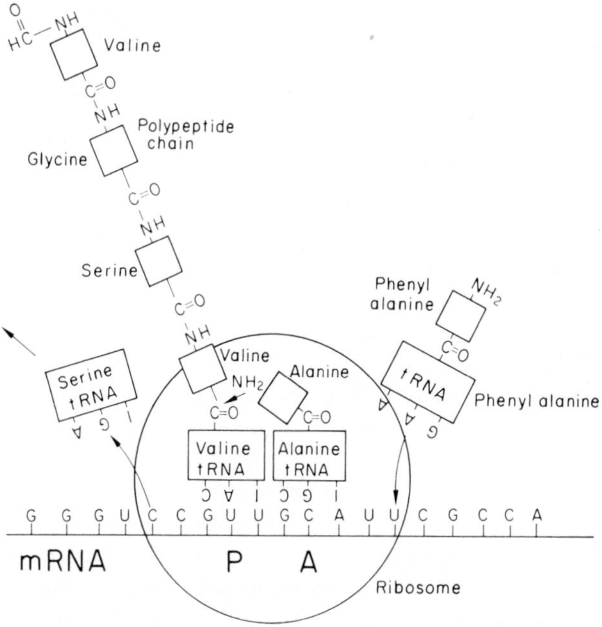

Fig. 19. The mode of elongation of a polypeptide chain. P is the "donor" or "peptidyl" site and A is the "acceptor" site. At the A site codon–anti-codon recognition occurs when the amino acyl tRNA enters the system. (After Crick[1].)

at the NH_2 terminal and proceeds in the direction of the carboxy-terminal. The stages of elongation may be stated as follows:

(i) The amino-acyl tRNA "recognizes" its place in the mRNA and a codon–anti-codon complex is formed. This step is aided by an *amino-acyl tRNA binding enzyme* termed T_1 and by GTP. Non-enzymatic codon–anti-codon recognition is also possible but this is only possible under high Mg concentrations in the order of 20 mM Mg^{++}. Under normal physiological conditions Mg^{++} concentrations are in the proximity of 5 mM thus necessitating T_1 intervention for binding. A peptide bond is formed between the carboxy (COOH) terminal group of the amino acid that has already been incorporated and the amino (NH_2) group of the incoming amino acid (see Fig. 19 and also Fig. 15). This stage, also powered by GTP, is catalysed by transfer enzymes called *peptidyl transferases* which are structural units of the larger ribosomal subunit (60 S or 50 S) located between the A and P sites.

(ii) After formation of the peptide bond the "old" tRNA sheds its amino acid and as schematically represented in Fig. 19 moves left. The enzyme functioning in this step is *translocase*, termed T_2 in plants[8] and probably identical to the bacterial G-factor. As a result of this shifting the discharged tRNA is released into the cytoplasm, the tRNA which previously carried amino acid moves from the "acceptor" to the "donor" site and a new amino-acyl tRNA molecule is attached at the acceptor site.

Termination

Three codons, UAA (ochre), UAG (amber) and UGA, do not promote the binding of amino-acyl tRNA. As regards UGA, this only applies to bacteria, while in higher organisms this is the codon for cysteine (see Table 2 on p. 39, the genetic code). These sequences code for termination of polypeptide chain elongation. Possessing this quality and the inability to code for amino acids, they are designated *nonsense codons*. Proof that they indeed do terminate elongation has been obtained by the isolation and characterization of the truncated polypeptide chains made by *in vivo* translation of messengers from genes containing nonsense mutations. It can also be shown that direction of translation from left to right (as shown schematically below) is sequential, and insertion of a termination codon at any point stops synthesis of gene products of genes to right, and not to the left, of the site of insertion. The disruption, by mutation, of sequential gene translation along a chromosome in a given direction is called the "polarity effect".

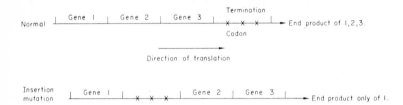

There is no common agreement as to why these codons cause termination. According to one opinion they cause ribosomes to "fall off" the mRNA while another states that they may prevent any further attachment of amino-acyl tRNA molecules to the polysomal apparatus. Recently it has also been shown that at least in *E. coli* termination is connected with specific "release factors" R_1, R_2 and S. R_1 causes termination by acting in conjunction with the UAA and UAG codons, while R_2 acts on UGA and UAA. After having shed its amino-acid load tRNA may recycle and participate in the process of translation several times before being broken down.

The ribosomal cycle. It is also believed that cycling of ribosomal subunits occurs during translation. The "ribosomal cycle" as seen in Fig. 20 indicates that after termination of the peptide chain elongation, the ribosome is released from the mRNA chain and thus becomes dissociated into its two subunits, 40 S and 60 S (or 30 S and 50 S). These subunits remain separate entities and only reassociate upon commencement of a new translational cycle. Like in the mazurka dance, partners are not necessarily kept and two given subunits do not as of need recombine.

In conclusion it must be remembered that several ribosomes may move along the chain at one and the same time thus producing several copies of the peptide.

Polysome assembly and seed dormancy. Initially observed during the process of fertilization of dormant sea urchin eggs, later demonstrated and detailed in Marcus' laboratory for eukaryotic wheat embryos (see for example, Marcus and Feely[9]) and subsequently in a wide variety of seeds upon release from dormancy, the phenomenon of polysome assembly now appears to be a critical factor in emergence of seeds from the dormant state. A typical and elaborate demonstration of the process in *Pinus resinosa* has recently been presented by Sasaki and Brown.[13]

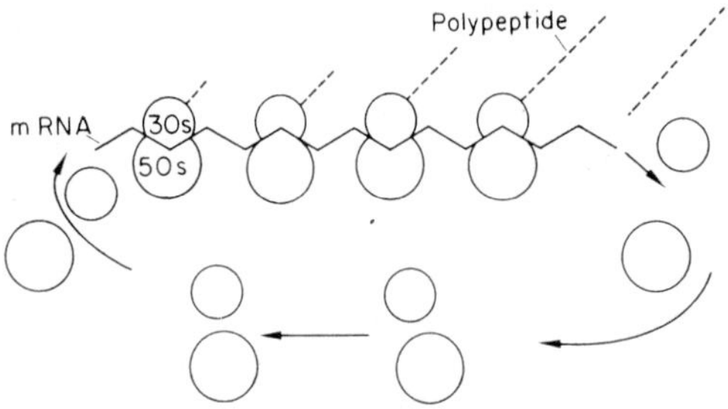

FIG. 20. The ribosomal cycle.

Using sucrose density gradients ribosomal preparations of dormant and imbibed embryos were layered over a 7·5–60 per cent (w/v) gradient, ultra-centrifuged at $110,000 \times g$ for 3 hours and subsequently fractionated and scanned at 253 nm. Results (see Fig. 21) indicate that during dormancy embryo ribosomes consist only of monomer units settling in one given region of the gradient. With progress of imbibition (the hydration of the dormant seed regarded as the initial step in seed germination) the profile changes and after 24 hours, instead of one monomeric peak, several others appear—these being considered polymers consisting of two or more ribosomes attached to mRNA. These polymers being larger molecules than the monosomes settle progressively at different peaks along the gradient and it was shown that with the progress of imbibition polysomes quantitively increased and monomer units decreased. A test often used to support evidence that the polysomal peaks are indeed what they are claimed to be is to incubate the preparation with RNase before layering over gradients. This enzyme, by hydrolysis of the mRNA strand linking the polymer, would cause its breakdown into monomeric units. A single peak in the monomeric region of the gradient produced by RNase incubation of the polymeric preparation (after 24 hours imbibition) would prove the point and this has been demonstrated by Sasaki and Brown (see Fig. 22). It is of interest to know whether bud dormancy, which in many aspects is similar to seed dormancy, follows the same pattern, and how certain plant hormones known to be active in dormancy break may affect the process.

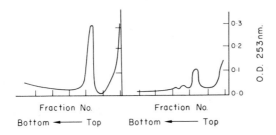

FIG. 21. Changes in sucrose density gradient ribosomal profiles of pine embryos during early stages of germination. *Left:* dormant embryos. *Right:* after 20 hours imbibition. (After Sasaki and Brown[13].)

Non-ribosomal protein synthesis. While discussing the mechanism of protein or oligo-peptide synthesis, the existence of a non-ribosomal system should be remembered. As pointed out by Lengyel,[3] the non-ribosomal mechanism permits synthesis of only short polypeptides and is far less specific than the ribosomal mechanism, since amino acids may in this case substitute for each other. Upon comparing both processes, the ribosomal and the non-ribosomal, it appears that the initial steps of both are similar. The amino-acid activation is the same, both processes require ATP and the result in either case is an amino acid enzymatically bound to adenylate (see p. 41). In both cases the polypeptide chain grows from the amino towards the carboxy terminal. However, the processes differ in that non-ribosomal synthesis, the acceptor of acti-vated amino acid, is not tRNA but a group of sulphydryl (SH) rich enzymes. The sequence of amino acids in the polypeptide chain is in this case determined by the sequence of the SH containing enzymes in the multi-enzyme complex which catalyses the reaction, and not a mRNA

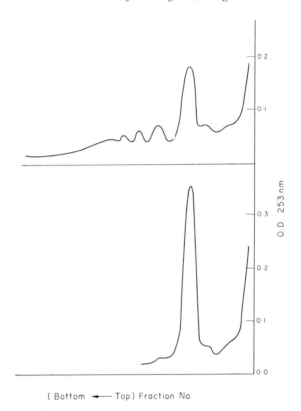

Fig. 22. *Above:* Sucrose density gradient profile of a polysomal preparation of embryos after 24 hours imbibition. *Below:* The same preparation after treatment with 1 μg/ml RNase for 5 minutes at 37°C. Note marked increase in monomer and disappearance of polymeric peaks. (After Sasaki and Brown[13].)

contained nucleic base sequence. As a result the size of the final product is limited by the size of the multi-enzyme complex.

Translational control of protein synthesis

In Chapter 4 we mentioned two control mechanisms of protein synthesis operating at a transcriptional level: (a) the classic operon model and its recent modifications and (b) the histone hypothesis. Now, after having outlined the process of translation, mention must be made of mechanisms operating at a translational level, which in higher plants may, together with histone control, be of prime importance. The major modes of translational control are:

1. *Ribosome-messenger attachment.* This mechanism may be either positive or negative. In the former case, specific factors may promote ribosome-messenger attachment and specific initiation factors might be required for the attachment of particular messengers to ribosomes. Such factors are analogous to the RNA polymerase associated factors functioning in transcription mentioned on p. 24. In the case of negative control, specific factors, possibly proteins, may prevent attachment.

2. *tRNA levels.* A hypothesis termed the "Adaptor Modification" hypothesis (Sueoka and Sueoka[14]) states that availability of tRNA may play a central role in protein synthesis control. A glance at the genetic code indicates that for most amino acids there are *several* codons. However, we have assumed that each molecule of tRNA contains only *one* anti-codon loop. The conclusion drawn from this is that "isoacceptor" tRNA molecules exist for such degenerate or redundant amino acids. Such isoacceptor tRNA's have been isolated in plants by Cherry. Under the "Adaptor Modification" hypothesis, it is assumed that for the synthesis of certain proteins, certain isoacceptors are preferred above others. Preferential availability of a given isoacceptor would therefore lead to preferential protein synthesis and lack of a certain type may shut off protein synthesis of a protein requiring this isoacceptor. The principle is illustrated in Table 4.

TABLE 4. SCHEMATIC REPRESENTATION OF THE "ADAPTOR MODIFICATION" HYPOTHESIS

(After Sueoka and Sueoka)[14]

Degenerate codons	Corresponding adaptors (or isoacceptor tRNA's)	Utilization of codons for protein synthesis		
		System 1	System 2	System 3
C_1	A_1	C_1	—	C_1
C_2	A_2	C_2	C_2	—
C_3	A_3	C_3	C_3	C_3
C_4	A_4	C_4	C_4	C_4

In this table C_1, C_2, C_3 and C_4 represent a set of degenerate codons for an amino acid and A_1, A_2, A_3 and A_4 are their corresponding isoacceptor tRNA molecules. System 1 utilizes all four codons for protein synthesis, system 2 does not use C_1, system 3 does not use C_2 and so on. If adaptor A_1 is lacking, protein synthesis in systems 1 and 3 is shut off without effecting system 2. Inactivation or lack of A_2 closes systems 1 and 2 but not system 3. If, on the other hand, only A_2, A_3 and A_4 are functional to start with, so that only system 2 is operating, the appearance of A_1 activates systems 1 and 3 without shutting off system 2. This mode of control is thus not only quantitative but also qualitative.

Evolution and the genetic apparatus

The discovery of the mechanism of DNA replication, transcription and translation into protein has provided further insight to biological development from an evolutionary point of view. Crick[1] and Orgel[11] have considered the process and related it to fundamentals of molecular biology such as the universality of the genetic code (see p. 38) and the primeval link between polypeptides and polynucleotides.

Orgel claims that amino acids, nucleotides and sugars could be formed from simpler molecules like formaldehyde (HCHO), hydrocyanic acid (HCN) and cyanoacetylene. Such processes according to him did actually take place in pre-biotic conditions and polypeptides and polynucleotides also made their appearance. According to his opinion, the genetic code, i.e. specific nucleotides for specific amino acids, was a matter of pure randomness. In other words, at the dawn of biological life on earth a crude correlation was established between certain polynucleotide sequences and certain available amino acids and the whole coding system "perfected itself by the bootstrap principle".

It is assumed that the non-biological polymerization of amino acids occurred before appearance of nucleic acids whose formation was probably biotic. These facts are not, however, the essence of self-replicating molecules in the "primeval soup" since possibly only the polypeptides were able to replicate themselves by the principle of attraction between positive and negative charges.

Nucleic acids as replication units. When simple derivatives of adenine and uracil are dissolved in aqueous medium, dimer crystals in which A and U are hydrogen bonded are obtained. The same applies to mixtures of guanine and cytosine. However, mixed crystals containing other pairs of bases are unknown. This implies base pairing as described in Chapter 2, and furthermore this is a phenomenon which is noted in non-biological conditions in aqueous media. In organic solvents this pairing is more pronounced and AU and GC pairs associate more strongly than any others, and when phosphorylated, helices form.

Orgel proposes that in a primeval state amino acids were loaded onto tRNA *not* by means of specific enzymes for each amino acid but directly. Crick[1] considers that this loading of amino acids might possibly have been by stereochemical fit between certain cavities on nucleic acids and the COOH terminal of the amino acid. This fit according to Crick may originally have been non-specific and enabled all available amino acids to affix themselves to tRNA, but during the evolutionary process a specific type of tRNA evolved for each amino acid. At this stage the process was "frozen" and an interaction equilibrium obtained between amino acids and tRNA's. If mutations not in accord with the evolved code did appear they were strongly selected against, since such changes would be highly disadvantageous as compared to the comparatively stable forms. Thus a mutation in a new direction would probably be lethal since only "nonsense" would result.

The development of specific enzymes—the aminoacyl-tRNA synthetases mentioned earlier in this chapter—according to Orgel would have been as follows. Suppose that CCC selects preferentially the amino-acid glycine and GGG preferentially selects proline; then if a polynucleotide CCC–GGG–GGG–GGG–CCC–GGG were formed, it would then tend to line up a polypeptide with a glycine–proline–proline–proline–glycine–proline sequence and this particular sequence may be a primitive activating enzyme causing somewhat more reliable association of CCC with glycine. Thus this activating enzyme would permit its own synthesis with greater reliability.

Upon being elongated the polynucleotide which is a long strand could double back on itself (see Fig. 23) by forming a loop and a double stranded base paired region beyond it. The region of looping probably is the single stranded loop region(s) recently determined for tRNA molecules (see Fig. 16). At this stage the reader is reminded of the fact

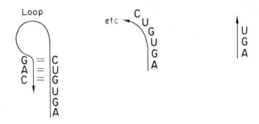

FIG. 23. The origin of tRNA according to Orgel.[11]

mentioned in Chapter 1, that Khorana's group has succeeded in the *in vitro* synthesis of alanine tRNA, a 77-nucleotide chain.

The above theories necessitate the existence of intermediary forms of life linking known and existing biological organisms and the above described primitive biotic ancestors. The search for such forms extending beyond the limitations of the terrestrial globe has as yet not produced positive results.

RECOMMENDED READING AND REFERENCES

1. CRICK, F. H. C., The origin of the genetic code. *J. Molec. Biol.*, **38**, 367–9 (1968). An article similar in nature to that of Orgel's (ref. 11) and stressing evolutionary aspects of the origin of the genetic code.
2. LENGYEL, P., Problems of protein biosynthesis. *J. Gen. Physiol.*, **49**, 3 (supp.), 305 (1966). An excellent article by one of the pioneers in the field. A succinct description and summary of the various factors participating in the process.
3. LENGYEL, P., The process of translation as seen in 1969. *Cold Spring Harb. Symp. Quant. Biol.*, **xxxiv**, 828–41 (1969). An updating of no. 2 with special reference to translation.
4. LIPMANN, F., Polypeptide chain elongation in protein biosynthesis. *Science*, **164**, 1024–31 (1969). A summary of the author's and other research on elongation including clearly stated descriptions of current theories and roles of enzymes, initiation and elongation factors.
5. LOENING, U., Molecular weights of ribosomal RNA in relation to evolution. *J. Molec. Biol.*, **38**, 355–65 (1968). Some thoughts on evolution in terms unknown to Darwin.
6. MARCKER, K. and SANGER, F., *N*-formyl methionyl s-RNA. *J. Molec. Biol.*, **8**, 835–40 (1969). Fundamental research on the "initiation" stage of protein synthesis.
7. MARCUS, A., Tobacco mosaic virus ribonucleic acid dependent amino acid incorporation in a wheat embryo system *in vitro*. Analysis of the rate of limiting factor. *J. Biol. Chem.*, **245**, 955–61 (1970). Work with plants indicating that "initiation factors" are not only found in bacteria.

8. MARCUS, A., Polypeptide synthesis in extracts of wheat germ. Resolution and partial purification of the soluble transfer factors. *J. Biol. Chem.*, **245**, 2814–18 (1970). As no. 7 above concerning transfer factors.

9. MARCUS, A. and FEELEY, J., Protein synthesis in imbibed seeds. II. Polysome formation during imbibition. *J. Biol. Chem.*, **240**, 1675–80 (1965). A now classic paper indicating that dormancy break is connected with the activation of mRNA by formation of a functional unit—the polysome.

10. NIRENBERG, M., The flow of information from gene to protein, in *Aspects of Protein Synthesis*, ed. C. B. ANFINSEN, Academic Press, 1970, 463 pp. Amongst other aspects of the process, a detailed description of polysome formation and function.

11. ORGEL, L. E., Evolution of the genetic apparatus. *J. Molec. Biol.*, **38**, 381–93 (1968). An article delving deeply into basic problems of evolution. A combination of metaphysics and molecular biology and an interesting discussion on how biological macromolecules could have been formed.

12. REVEL, M., BRAWERMAN, G., LELONG, J. and GROS, F., Function of three protein factors and ribosomal subunits in the initiation of protein synthesis in *E. coli. Nature*, **219**, 1016–21 (1968). An important contribution to the understanding of polypeptide synthesis. Detailed experimentation on "initiation" factors.

13. SASAKI, S. and BROWN, G. N., Polysome formation in *Pinus resinosa* at initiation of seed germination. *Pl. Cell Physiol.*, **12**, 749–58 (1971). An elaborate demonstration that emergence from the dormant state in pine seeds is related to shift of ribosomal monomers to polymers.

14. SUEOKA, N. and SUEOKA, T. K., Transfer RNA and cell differentiation, pp. 23–55 in *Prog. Nucl. Acid. Res. and Molec. Biol.*, **10**, ed. J. DAVIDSON and W. COHN, Academic Press, 1970. A clear indication that in addition to "operon" control of protein synthesis other mechanisms exist, including control at the level of translation.

In spite of the extensive chloroplastic protein synthesis the process of replication is not entirely independent of nuclear control and Brawerman and Chargaff[3] working on *Euglena* have reported "an autonomous replication factor" formed in the nucleus and necessary for chloroplast multiplication. DuPraw[4] has claimed that chloroplastic DNA is responsible only for the synthesis of chloroplastic ribosomes which combine with mRNA of nuclear origin to form protein synthesizing polysomal chloroplastic units. Kirk[5] has demonstrated the contribution of nuclear DNA to some of the central functions of the chloroplast, the formation of the photosynthetic pigments and of the enzymes participating in electron transport.

RECOMMENDED READING AND REFERENCES

1. ATTARDI, G., ALONI, Y., ATTARDI, B., OJALA, D., PICA-MATOCCIA, L., ROBBERSON, D. and STORRIE, B., The transcription of mitochondrial DNA in Hela cells. *Cold Spring Harb. Symp. Quant. Biol.*, xxxv, 599–619 (1970). An up-to-date account of the process by one of the foremost research groups on mitochondrial mechanisms.

2. BORST, P. and KROON, A. M., Mitochondrial DNA, in *Int. Rev. Cytol.*, pp. 107–90, ed. E. BOURNE and H. DANIELLI, Academic Press, 1969. An extensive review and detailed descriptions of all known types of mitochondrial DNA. A special section is devoted to plant mitochondria.

3. BRAWERMAN, G. and CHARGAFF, E., A self-reproduction system concerned with the formation of chloroplasts in *Euglena gracilis*. *Biochem. Biophys. Acta*, **37**, 221–9 (1960). Research indicating that replication of chloroplastic DNA depends on nuclear factors as well.

4. DuPRAW, E. J. See reference 4 in Chapter 2.

5. KIRK, J. T. O., *Biochemistry of Chloroplasts* **1**, 319, ed. T. W. GOODWIN, Academic Press, N.Y., 1966. Determination of which chloroplastic constituents are coded for by chloroplast DNA and which by nuclear DNA.

6. NASS, M., Mitochondrial DNA: advances, problems and goals. *Science*, **165**, 25–35 (1969). An objective and clear review providing the different points of view concerning origin and development of mitochondria.

7. RAVEN, P., A multiple origin for plastids and mitochondria. *Science*, **169**, 641–6 (1970). The author, a protagonist of the symbiont concept of organelle evolution, discusses his viewpoint.

8. STOCKING, C. R. and GIFFORD, E. M., Incorporation of thymidine into chloroplasts of *Spirogyra*. *Biochem. Biophys. Res. Commun.*, **1**, 159–64 (1959). One of the first steps which indicated partial autonomy of the chloroplast as a genetic unit.

9. WILKIE, D., Reproduction of mitochondria and chloroplasts, in *Organization and Control in Prokaryotic and Eukaryotic Cells*, pp. 381–99, ed. H. CHARLES and B. KNIGHT, Cambridge University Press, 1970. A review of the topic from several aspects: biochemical, genetic and evolutionary. A wealth of information condensed in a series of review articles.

CHAPTER 8

Antibiotics and Inhibition of Synthesis
of Nucleic Acids and Protein

IN 1961 Reich *et al.*[10] reported that minute amounts of the antibiotic actinomycin, when applied to a tissue culture, inhibited RNA synthesis while synthesis of DNA and protein, at least initially, was not affected. This discovery proved of great interest to molecular biologists since it provided a means whereby a specifically acting inhibitor could be used in order to study the process of protein synthesis. Since the above report many other antibiotic substances having specific inhibitory action on various stages of protein synthesis have been discovered. As shall be explained shortly, the inhibition may be on enzyme systems and in several cases may be caused by structural similarities between the normal constituents and the antibiotic. Cases of inhibition at a template level are also known. Hartman *et al.*[6] have attempted to review antibiotic action systematically and their classification, to a major degree, is followed in the present text.

INHIBITION OF NUCLEIC ACID SYNTHESIS

(a) *Effect on RNA polymerase*

As outlined in Chapter 4 the initial step of transfer of genetic information contained in a nucleotide sequence on DNA is the formation of a strand of RNA. The nucleotide triphosphates under polymerization mediated by the enzyme RNA polymerase. This reaction may be summarized as follows:

$$n_1 ATP$$
$$n_2 UTP \xrightarrow[\text{RNA polymerase}]{\text{Mg}^{++}\text{DNA}} (n_1 + n_2 + n_3 + n_4) \, PP + RNA$$
$$n_3 GTP$$
$$n_4 CTP$$

In order to understand the effects of a number of antibiotics the mechanism of RNA polymerase action shall be again detailed briefly. As seen in Figure 25, the synthesis of RNA proceeds in several stages. Initially the polymerase attaches itself to the DNA. In the event that the DNA is double stranded as in higher organisms it is believed that the DNA is opened or "melts" at the point of RNA polymerase attachment and only afterwards does polymerization of nucleotides occur. The polymerase travels along the strand synthesizing RNA as it proceeds and inorganic pyrophosphate is freed. Under certain conditions, mentioned previously, the synthesized RNA is set free and it is possible that the rate of RNA release is the limiting factor of the process in its entirety.

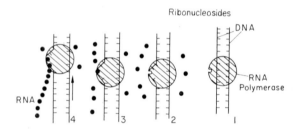

FIG. 25. Synthesis of DNA directed RNA by RNA polymerase. (After Hartman *et al.*[6])

Rifamycin and rifampicin. These two substances (see Fig. 26), the latter being the hydrazone derivative of the former, are claimed to be specific RNA polymerase inhibitors. In the bacterium *E. coli* a concentration of only 3×10^{-8} M of rifamycin caused a 50 per cent inhibition of the enzyme. It is commonly held opinion that the antibiotic is specific to bacteria since mouse RNA polymerase was not affected. However, in plants, evidence exists that at least in certain families it has activity.

Rifamycin and rifampicin only inhibit the free enzyme before its attachment to DNA (stage 2 in Fig. 25), and if this has already occurred and if the nucleotides have been added (stage 3 in Fig. 25) the enzyme is protected against the action of these antibiotics. These findings therefore indicate that their inhibition is at a very early stage in the process of

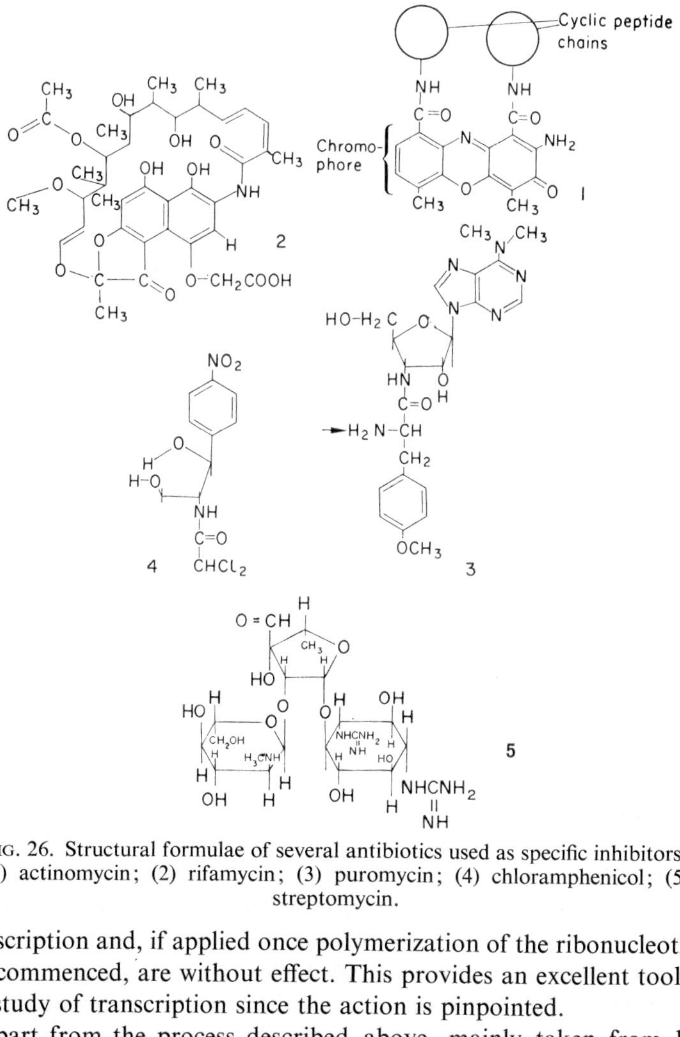

FIG. 26. Structural formulae of several antibiotics used as specific inhibitors:
(1) actinomycin; (2) rifamycin; (3) puromycin; (4) chloramphenicol; (5)
streptomycin.

transcription and, if applied once polymerization of the ribonucleotides
has commenced, are without effect. This provides an excellent tool for
the study of transcription since the action is pinpointed.

Apart from the process described above, mainly taken from bacterial systems, rifampicin is unique as an antibiotic in that effects on
viruses have also been found (Subak-Sharpe *et al.*;[11] Becker and
Zachay-Rones[2]). It has been reported effective against the vaccinia,
smallpox and trachoma viruses. The latter virus, causing eye infection

that this antibiotic also interferes with the attachment of amino-acyl tRNA molecules to the A or P sites have also been presented. Figures 27 and 28 are a schematic representation of the above-mentioned two modes of action. Figure 27 indicates how streptomycin may cause a change on the 30 S ribosomal subunit resulting in misreading of the codon, the attachment of the "wrong" tRNA molecule and finally the incorporation into the polypeptide chain of an amino acid not contained in the codon. This figure indicates how the UUU codon should accept the phenyl-alanine tRNA (right) but upon action of streptomycin the UUU is read as if it were AUU, thus causing incorporation of iso-leucine (left).

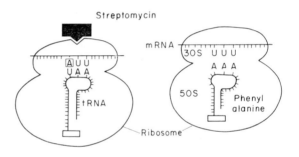

FIG. 27. The action of streptomycin in causing incorrect reading of the codon. (After Gorini[4].)

Figure 28 indicates how the antibiotic may correct two mutations which ordinarily would cause synthesis of non-functional or unstable proteins. In *nonsense* mutations (Fig. 28 below) the correct codon, for example, CAG (right), undergoes a mutation and the resulting codon UAG is a triplet which codes for no amino acid, thus preventing attachment of any amino-acyl tRNA whatsoever. Streptomycin may correct this defect by causing the "nonsense" to be read as if it were sense and that nothing had occurred. As a result the codon, despite the nonsense mutation UAG, is read correctly and glutamine-tRNA containing the anti-codon GUC to the correct codon CAG is attached.

Figure 28 below indicates a *missense* mutation which differs from a nonsense one. In this case the correct codon, e.g. CCU which codes for proline (right), is mutated to CAU which codes for histidine (middle). Streptomycin may cause the CAU to be read as if it were CCU and thus, despite the missense, which ordinarily would result in a defective end-product, it is read correctly and proline incorporated (right).

This antibiotic is specific for 70 S ribosomes and has no effect on eukaryotic cytoplasmic 80 S ribosomes. For this reason it has been used extensively in studies on chloroplasts possessing the lighter unit.

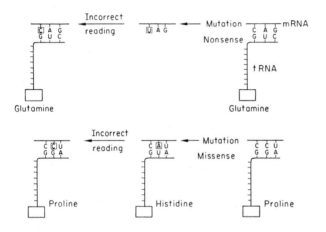

Fig. 28. The correction of "nonsense" (above) and "missense" mutations by streptomycin. (After Gorini[4].)

Chloramphenicol. This substance (Fig. 26) in general acts similarly to the previously described one but with the difference that it attaches itself to the 50 S subunit and not the 30 S. Here, too, the antibiotic is specific for 70 S ribosomes. Its exact mode of protein synthesis inhibition has as yet not been elucidated but there are indications that it inhibits the action of the enzyme peptidyl transferase.

Puromycin. A glance at the structure of puromycin (Fig. 26) reveals that this antibiotic is a nucleotide. Its mode of action as a protein synthesis inhibitor is by its amide linkage between the amino group of its methyl tyrosine "tail" (indicated by an arrow in Fig. 26) and the carboxy-terminal of the elongating polypeptide chain (Nathans[9]). This mode of action is apparent upon considering the resemblance of the puromycin molecule and the 3′ end of amino-acyl tRNA.

As a result of this binding which occurs at the A site on the ribosome and as a result of its inactivity yet resemblance to amino acid carrying tRNA, the puromycin essentially assumes the role of a competitor of the enzyme peptidyl transferase. It is incorporated to the existing poly-peptide chain but since no further amino acids may be incorporated it causes release of the chain from the ribosome and cessation of protein synthesis. The mode of action also explains experimental observations that the application of puromycin results in the release of short peptide chains. Puromycin action is broad, extending to both eukaryotic and prokaryotic ribosomes.

It is not the purpose of this chapter to review all of the major anti-biotics used in research, but rather to provide a general framework stating several typical examples. When considering effects of antibiotics it must be kept in mind that mechanisms of inhibition discovered and worked out under *in vitro* conditions may be a far call indeed from effect *in vivo*. In the latter case the conditions and dosages applied under experimental conditions would be lethal and furthermore, in the com-plete organism, beside the genetic apparatus there are several other possible sites of interaction. These may include preformed enzymes, membrane systems, etc., and moreover it is known that membranes of certain organelles or even whole cells are impermeable to certain kinds of antibiotics.

RECOMMENDED READING AND REFERENCES

1. BECKER, Y., ASHER, Y., HIMMEL, N., ZACHAY-RONES, Z. and MATHYAR, B., Rifampicin inhibition of *Trachoma* agent *in vivo*. *Nature*, **224**, 32–34 (1969). This research together with that quoted below (nos. 2, 3, 7 and 11) are a "break-through" in the study of antibiotics. An indication that antibiotics in addition to being anti-bacterial may also be anti-viral.
2. BECKER, Y. and ZACHAY-RONES, A., Rifampicin—a new antitrachoma drug. *Nature*, **222**, 851–3 (1969). See no. 1 above.
3. BEN-ISHAI, Z., HELLER, E., GOLDBLUM, N. and BECKER, Y., Rifampicin, *Poxvirus* and *Trachoma* agent. *Nature*, **224**, 29–33 (1969). See no. 1 above.
4. GORINI, L., Antibiotics and the genetic code. *Scient. Am.*, **214**, 102–9 (1966). A semi-popular article explaining and illustrating how antibiotics, especially streptomycin, can cause genetic changes in cells.

5. GORINI, L. and KATAJA, E., Phenotypic repair by streptomycin of defective genotypes in *E. coli. Proc. Nat. Acad. Sci.*, **51**, 3, 487–93 (1964). Antibiotics do not always have negative effects. A research describing "correction" of genetic defects by streptomycin.

6. HARTMAN, G., BEHR, W., BEISSNER, K. A. and SIPPEL, A., Antibiotics as inhibitors of nucleic acid and protein synthesis. *Angew. Chem., Int. English Ed.*, **7**, 693–701 (1968). A thorough review of effects of the major classes of antibiotics and detailed discussion on their mode of action.

7. HELLER, E., ARGAMAN, M., LEVY, A. and GOLDBLUM, N. Selective inhibition of *Vaccinia* virus by the antibiotic rifampicin. *Nature*, **222**, 273–4 (1969). See no. 1 above.

8. MULLER, W. and CROTHERS, D., Studies of the binding of actinomycin and related compounds to DNA. *J. Molec. Biol.*, **35**, 251–90 (1968). Actinomycin is known to cause general inhibition. Experiments indicating how it binds directly to DNA and how blockage is caused.

9. NATHANS, P., Puromycin inhibition of protein synthesis: incorporation of puromycin into peptide chains. *Proc. Nat. Acad. Sci.*, **51**, 585–92 (1964). An exact biochemical research pinpointing inhibition of protein synthesis as caused by puromycin.

10. REICH, E., FRANKLIN, R., SHATKIN, A. and TATUM, E., Effect of actinomycin D on cellular nucleic acid synthesis and virus production. *Science*, **134**, 556–67 (1961). Pioneering work opening the way to understanding the connection between action of antibiotics and the protein synthesizing apparatus.

11. SUBAK-SHARPE, J. H., TIMBURY, M. C. and WILLIAMS, J. F., Rifampicin inhibits the growth of some mammalian viruses. *Nature*, **222**, 341–5 (1969). See no. 1 above.

the chances of integration of such transducing particles into host chromosomes are 1:10.

Another and somewhat different instance of transduction pertains to the λ-phage as described by Thomas.[8] This phage infects bacteria, is lysogenically integrated into host DNA and replicates together with the host genome. It must be remembered that the site of insertion of the phage DNA into host DNA is specific and determined by a corresponding site on the phage producing a specific integrase enzyme. Other varieties of the same phage are integrated at different sites since they appear to possess different integrases.

After the integration of phage DNA the bacterium may set it free once again—the process being called "induction". Upon release the phage may detach a small segment of host DNA as well. Phage particles of this sort containing additional DNA can be isolated from a population of normal phages and it is possible to use them to reinfect normal cells which now, in addition to viral genome, also integrate additional host chromosomal segments containing several genes.

In both the above-mentioned examples, P_1 and λ, the final product is a virus mediated addition of genetic material to the host genome. In the first case, P_1, the addition is completely random while in the latter, λ, it is manifested as supplementation of given areas of "donor" host DNA areas which are adjacent to the specific site of initial integration of the phage DNA.

Supplementary genetic material may exist independently of the host DNA in the form of non-nuclear particles which, if non-obligatory and extra elements, are termed *episomes*. Episomes can exist autonomously but can integrate into the chromosome. Obligatory non-nuclear genetic elements are termed *plasmids* and under this heading organelle (mitochondrial and possibly chloroplast) DNA may be included.

Experimentation as reported above has primarily been carried out on bacterial cells while in eukaryotic systems only partial success has been reported. Dulbecco[3] and also Stein and Enders[7] have demonstrated that when mammalian cells are infected with certain viruses, e.g. polyoma virus, simian virus 40 or adeno-virus, a certain number of virus particles undergo lysogenic transformation. However, it is not quite clear if actual insertion takes place or whether the viral DNA exists as episomes. In either event the supplemented genetic material is replicated in the same manner as nuclear DNA and new types of mRNA, protein and, as a result, also antigenic material make their appearance.

To indicate the therapeutic potential of transduction, the example recently reported by Schwartz *et al.*[5] shall be stated in brief: a known genetic defect is the lack of ability to produce inosinic acid pyrophosphorylase (IAP). Mouse cells without this enzyme are unable to develop in certain media. Using an inactivated mouse virus small amounts of healthy genetic material from chick embryos were inserted into mouse-cell cultures in a culture medium in which IAP-deficient cells do not survive. After insertion of the foreign genetic material most mouse cells died. However, some clones appeared within 2 to 3 weeks after cell fusion. These mouse-cell clones retained the chick gene for the enzyme and the gene was continually replicated on cultivation of the cells *in vitro*.

A problem posed by these workers is that in addition to genes, immunological antigenic material is also introduced and these antigens could initiate a homograft rejection reaction. They claim that the obstacle may be overcome by using amounts of genetic material which are too small to code for the antigens.

The above approach holds great promise but practical application is still in its infancy. The probability of achieving exact results from transduction of given mapped genes is greater in bacteria where only one chromosome exists. In higher organisms such as mammals and plants the lysogenic virus has the choice of many chromosomes which more often than not are unmapped and the chances that genes from normal cells correspond to the genetic defective ones of the "acceptors" are slim indeed. In spite of these and other difficulties, scientists do not consider this an unsurmountable task, and research on these lines is being continued. Sinsheimer[6] has postulated that this technique may one day solve the malady of diabetes which can be caused by the lack of a certain known protein—proinsulin. A lysogenic virus carrying the proinsulin gene section of the chromosome of normal healthy "donor" cells could theoretically be inserted into pancreatic cells of a diabetic patient and restore the genetic ability to synthesis proinsulin.

RECOMMENDED READING AND REFERENCES

1. CAMPBELL, A., Episomes. *Advances in Genetics*, **11**, 101–45 (1962). A presentation of the fundamental concept of mode of lysogenic virus "invasion" and integration into host chromosomes.
2. DULBECCO, R., The state of the DNA of *polyma* virus and SV 40 in transformed cells. *Cold Spring Harbor Symp. Quant. Biol.*, **xxxiii**, 777–83 (1968). See no. 3 below.
3. DULBECCO, R., Cell transformations by viruses. *Science*, **166**, 962–8 (1969). A description of the process mainly in animal cells.
4. IKEDA, H. and TOMIZAWA, J., Transducing fragments in generalized transduction by phage Pl. *J. Molec. Bot.*, **14**, 85–109 (1965). Experimentation with "genetic engineering".
5. SCHWARTZ, A., COOK, P. and HARRIS, H. Correction of a genetic defect in a mammalian cell. *Nature N.B.*, **230**, 5–8 (1971). A glance at things-to-come? The correction of a genetic defect in mouse cells and a level-headed discussion of difficulties encountered.
6. SINSHEIMER, R. L. S. (1969) See reference 4 in Chapter 1.
7. STEIN, H. and ENDERS, J. Transformation induced by simian virus 40 in human renal cultures. I. Morphology and growth characteristics. *Proc. Nat. Acad. Sci.*, **48**, 1164–72 (1962). Research indicating that lysogeny and transfer of genetic material is not limited to bacteria.
8. THOMAS, R., Lysogeny, in *The Molecular Biology of Viruses*, pp. 315–42, 8th Symp. Soc. Gen. Microbiol., 1968, ed. L. CRAWFORD and M. STOKER. A detailed description of the process with stress on the infection stage.

CHAPTER 10

The Use of RNA for Transmission of Genetic Information

THE possibility of transmitting information using RNA has been studied under experimental conditions in whole organisms, tissue cultures and cell-free systems. This mode of informational transfer differs from transformation and transduction which essentially are DNA-mediated processes.

Much attention has been drawn to potential RNA-mediated informational transfer by the researches of neurobiologists such as Babich et al.[1] and Jacobsen et al.[2] Their work concentrated on the possibility of transfer of "memory" from mice which had learned to find their way through mazes to "unlearned" mice. In the reported experiments, the "learned" mice were killed, RNA extracted from their brains and injected into mice which had not been introduced into the experimental mazes. It was claimed that upon doing so, the unlearned individuals took a shorter time to find their way than an uninjected control group. Lutges et al.[9] attempted to repeat these experiments paying particular attention to RNA-extraction procedures and the prevention of RNA breakdown during the process and were unable to induce learning effects. It here appears that "memory" is an intricate and rather complex mechanism perhaps depending on a large number of interacting factors and hence the uncertainty of conclusions either positive or negative.

In more limited systems in which the final product is one protein or a defined enzyme, more specific research has been carried out and utilizing RNA as the source of information the goal seems to have been achieved. Proteins formed in this manner include haemoglobin (Chantrenne et al.[3]), serum albumin, tryptophane pyrrolase, glucose-6-

79

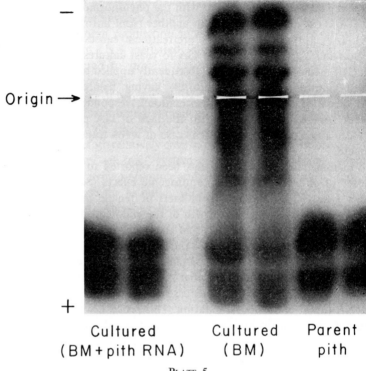

Origin →

+

Cultured   Cultured   Parent
(BM + pith RNA)   (BM)   pith

PLATE 5

RNA-mediated transfer of genetic information in tobacco pith culture. Expressed in peroxidase isozyme patterns as determined by starch-gel electrophoresis. *Right:* Isozyme pattern of parent pith. *Centre:* Changes of pattern experienced upon growing pith in basal culture medium (BM). *Left:* Pith which had been grown in culture medium and into which RNA from "parent" tissues had been infiltrated. (From Leshem and Galston[8].)

(Galston *et al.*[6]) that IAA prevents the formation of these additional isozymes. Leshem and Galston[8] extracted RNA from regular "parent" tissue and either injected or vacuum infiltrated the RNA into tissue which was transferred to culture media, the result being the prevention of appearance of the new isozymes which ordinarily appear in culture (Plate 5). A variation of this experiment was to extract RNA from tobacco pith tissue which had grown in IAA containing media, i.e. tissue in which the appearance of the new isozymes is prevented by IAA.

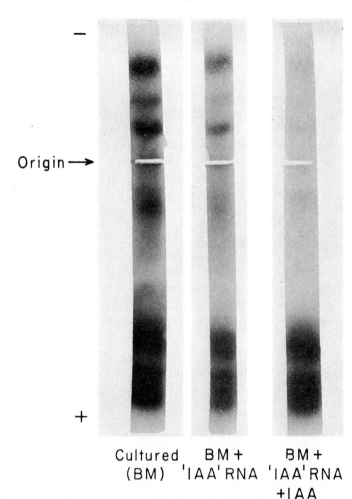

PLATE 6

Left: Peroxidase isozyme pattern of tobacco pith growing in basal culture medium (BM). *Centre:* Isozyme pattern of pith infiltrated with RNA extracted from an IAA-treated pith culture. *Right:* Addition of IAA to the "centre" system causes a significant fade-out of isozymes which ordinarily appear under conditions of culture. (From Leshem and Galston[8].)

This RNA was in turn applied as above to a second series of tissue in culture. The result, seen in Plate 6, again indicates that RNA mediated repression of isozyme appearance was obtained. This research also attempted to characterize the RNA and most activity was detected in a fraction having messenger-like properties. RNase or 0.3 N KOH hydrolysis of the isolated RNA markedly decreased its repressive effect.

A third group of exogenous RNA-elicited changes is the induction of immunological responses such as antibody production and homograft immunity. Reports have been made of RNA which was extracted from tissues possessing the ability to produce certain antibodies and which was able to transfer this ability to tissues normally incapable of synthesizing such antibodies (Fishman and Adler[5] and Bishop *et al.*[2]). Doubts have, however, been expressed concerning the validity of the conclusions drawn, since it may be contended that together with the applied RNA minute amounts of antigenic protein may also have entered the recipient cells thus evoking the specific antibody production. This contention has been overruled by the demonstration that pretreatment of the extract with RNase which hydrolyses the RNA and not the suspected protein, causes a nullification of the transferred immunoresponse thus indicating that the effect was elicited by the RNA itself and not antigen fragments.

RECOMMENDED READING AND REFERENCES

1. BABICH, F., JACOBSEN, A. L., BUBASH, S. and JACOBSEN, A., Transfer of a response to naïve rats by injection of ribonucleic acid extracted from trained rats. *Science*, **149**, 656–7 (1965). A pioneering effect with great promise. RNA-mediated transfer of "memory" from trained to untrained mice.
2. BISHOP, D. L., PISCIOTTA, A. V. and ABRAMOFF, P., Synthesis of normal and "immunogenic RNA" in peritoneal macrophage cells. *J. Immun.*, **99**, 751–9 (1967). The transfer of immunological effects from individual to individual.
3. CHANTRENNE, H., BURNY, A. and MARBAIX, G., The search for the messenger RNA of hemoglobin. *Prog. Nucl. Ac. Res. and Molec. Biol.*, **7**, 173–94, ed. J. DAVIDSON and W. COHN (1967). A dramatic description of one of the most intense searches conducted by molecular biologists. How globin messenger was found and tested.
4. FUJII, F. and VILLEE, C. A., Partial characterization of the RNA stimulating growth of the seminal vesicle. *Proc. Nat. Acad. Sci.*, **62**, 836–43 (1969). RNA-mediated hormonal informational transfer and characterization of the active RNA component.
5. FISHMAN, M. and ADLER, F. L., Antibody formation initiated *in vitro*. II. *Exp. Med.*, **117**, 595–602 (1963). As no. 2 above.

6. GALSTON, A. W., LAVEE, S. and SIEGEL, B. Z., The induction and repression of peroxidase isozymes by 3-indoleacetic acid, in *Biochemistry and Physiology of Plant Growth Substances*, pp. 455–72, ed. F. WIGHTMAN and G. SETTERFIELD. Runge Press, Ottawa, 1969. A detailed description of the experimental system subsequently used (see no. 8 below) for RNA experimentation.

7. JACOBSEN, A. L., BABICH, F., BUBASH, S. and JACOBSEN, A., Differential approach tendencies produced by injection of RNA from trained rats. *Science*, 150, 636–7 (1965). See no. 1 above.

8. LESHEM, Y. and GALSTON, A. W., Repression of isoperoxidase formation in excised tobacco pith by exogenous auxin-controlled RNA. *Phytochem.*, 10, 2869–78 (1971). RNA extracted from tobacco tissue cultures containing "peroxidase information" imposes its pattern on cultures lacking this information.

9. LUTGES, M., JOHNSON, T., BUCK, C., HOLLAND, J. and McGRAUGH, J., An examination of transfer of learning by nucleic acid. *Science*, 151, 834–7 (1966). An interesting challenge to memory-transfer phenomena. A repetition of the mouse "maze" experiment failing to verify previously reported success.

10. MASUDA, Y. and YANAGISHIMA, N., RNA functional in auxin action of expanding tuber tissue of Jerusalem artichoke. *Pl. Cell Physiol.*, 5, 365–8 (1964). One of the first experiments to be performed on plants with the aim of transferring hormone effects by nucleic acids.

11. MANSOUR, A. M. and NIU, A., Functional studies with uterine RNA. *Proc. Nat. Acad. Sci.*, 53, 764–70 (1965). The transfer of enzyme-induced effects.

12. NATHANS, D., NOTANI, G., SCHWARTZ, J. H. and ZINDER, N. D., Bio-synthesis of the coat protein of coliphage f_2 by *E. coli* extracts. *Proc. Nat. Acad. Sci.*, 48, 1424–31 (1962). A research of considerable theoretical value. It was shown that protein synthesis could take place in a heterologous cell-free system.

13. SEGAL, S. J., Regulatory action of estrogenic hormones, in *Control Mechanisms in Developmental Processes*, Supplement 1, pp. 264–80, ed. E. LOCKE, Academic Press, 1967. A general review discussing the transfer of estrogen and testosterone-induced effects without estrogen and testosterone.

14. ZOLOTOV, Z. and LESHEM, Y., Promotion of α-amylase production of isolated barley aluerones by RNA extracted from germinating embryos. *Pl. Cell Physiol.*, 9, 831–2 (1968). RNA from germinating barley embryos caused gibberellin-like effects under *in vitro* conditions.

CHAPTER 11

Interferon

IN 1957 two English scientists, Isaacs and Lindenmann,[3] characterized a protein, termed *interferon*, able to interfere with virus replication. This substance is produced in the cell upon infection with certain viruses and its presence inhibits development of a wide spectrum of non-related viruses and hence the assumption that its action has no connection with immunological phenomena. Interferon is species specific, i.e. it confers virus immunity only in the particular species infected by the virus and thus immunity cannot be transferred to any other species. To be more explicit—interferon produced upon viral infection, for example in horses, is ineffective in man infected with the same or other virus.

Physical properties of interferon include resistance to RNase and DNase, sensitivity to trypsin hydrolysis, and stability over a wide range of pH, from 2 to 10. The molecule is non-dialysable and depending upon species has a molecular weight from 25,000 to 150,000. As yet no report has been made of chemical isolation and characterization of interferon, but scientists have made considerable progress towards its purification and its existence is now believed to be more than hypothetical. It has also been demonstrated that antibiotics like actinomycin D and puromycin prevent its appearance under circumstances that normally would lead to its synthesis.

Mode of interferon action

It is an accepted fact that interferon as such has no inhibitive action on viruses since if added directly to viral cultures it is ineffective. It is believed that it acts in the cell and causes certain changes which enable the cell to check viral replication. It has also been demonstrated that interferon is non-toxic to cells.

The most efficient inducers of interferon production are viruses whose genetic material is double-stranded RNA, e.g. rheovirus 3, but instances have been documented whereby interferon production is also elicited by DNA viruses. The protein sheaths of the viruses cause a lag in interferon production since in order that the cell may react to viral infection these protein sheaths have to be shed. Application of "unsheathed" virus results in a far more rapid cell response.

It has been shown that interferon production can be induced by synthetic means, e.g. by the use of synthetic double-stranded RNA, one strand being a polymer of the nucleic base inosine and the other a polymer of cytidine. The double-stranded complex is known as poly I:C.[5]

One of the hypotheses as to the mode of action of interferon suggested by Marcus and Salb[4] is that as a result of viral infection the host cell ultimately produces a "translation inhibiting protein". This protein attaches to ribosomes and the polysomal units assembled in the infected host cell resultingly differ somewhat from the polysomes in the healthy cell. The altered polysomes only translate mRNA of the infected host cells and not of the invading RNA. Hilleman[2] outlines the following stages of interferon-induced viral inhibition:

Stage 1. The virus adheres to the cell, penetrates it and by shedding its protein sheath releases the single strand of contained RNA. The single strand is replicated and a RNA duplex formed. This elicits an "alert reaction" in the host cell expressed as localized derepression of a specific DNA segment which codes for interferon. Transcription and translation proceed and the interferon makes its appearance in the host's cytoplasm.

Stage 2. The interferon penetrates into neighbouring cells, there inducing specific transcription and translation which result in the synthesis of the previously mentioned protein which inhibits translation of viral RNA but has no adverse effects on the normal translational processes of the host cell.

Whether or not the above hypothesis is correct, the inhibitory effect of interferon has been encountered for a rather wide variety of viruses and attempts have been made to harness this substance in mankind's battle against both animal and plant viral infection. Experimentation has been conducted on mice infected with pneumonia virus. Ninety per cent of the experimental animals infected with lethal doses of the virus recuperated if pretreated with interferon-inducing RNA. All non-inoculated mice died. Park and Baron[5] have also reported prevention and even therapy of viral eye infection in rabbits.

In the above experiments interferon was induced by application of poly I:C mentioned above. Rabbits were treated as follows: one group was treated with poly I:C shortly before inoculation with *Herpes*

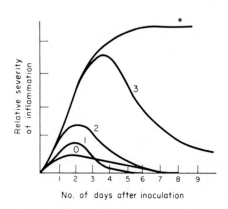

No. of days after inoculation

Fig. 30. Response of a virus causing rabbit-eye inflammation to treatment with double-stranded RNA-inducing interferon. (After Park and Baron.[5]) 0 = interferon inducer applied at the same time as inoculation with virus. 1 = interferon inducer applied 1 day after inoculation with virus. 2 = interferon inducer applied 2 days after inoculation with virus. 3 = interferon inducer applied 3 days after inoculation with virus. * = control—only inoculation with virus.

simplex virus which causes severe eye infection. Three other groups were treated 1, 2 or 3 days after virus inoculation. It was found (see Fig. 30) that if the synthetic RNA was applied at "zero" time, i.e. on the day of infection, only slight inflammation results. However, if applied 3 days after inoculation inflammation is severe but nevertheless is almost completely cured later. Reports have been made in the U.S.S.R. of attempts to use interferon inducers to combat the common cold, the cause of which is viral, but as yet no conclusive results have been produced.

Since many plant viruses are RNA viruses it is of interest to ascertain whether the interferon mechanism exists in the plant world. Virus-resistance phenomena possibly, but not conclusively, attributed to interferon have been found by plant virologists but the general opinion is that the definite existence of plant interferon still has to be demonstrated. An important milestone along the path of plant research in this field is the report of Stein and Loebenstein[6] who showed that poly I:poly C when injected into tobacco or *Datura* plants induced immunity to tobacco mosaic virus.

Stern[7] has recently reported that double-stranded RNA exists in normal cells either as a regular cell component or a latent virus. He also isolated a hitherto unknown enzyme which breaks down double-stranded RNA, including poly I:C. This enzyme promises to be an important tool in interferon research, since with its help dosage of poly I:C as a drug to fight virus can be determined more exactly. Another approach is to find a specific inhibitor of the enzyme (possible by antigenic methods) and thus increase the longevity of the applied drug.

Many laboratories, including those of pharmaceutical firms, are conducting intensive research in this field with the aim of controlling viral infections. There are, however, several obstacles which have to be overcome. These include the species specificity outlined in the beginning of this chapter, extremely localized effects, instability and the difficulty of inducing production of large amounts of the substance.

RECOMMENDED READING AND REFERENCES

1. FINTER, N. B., *Interferons*, North Holland, Amsterdam, 1966. A general outline of the subject.
2. HILLEMAN, M. R., Toward control of viral infection of man. *Science*, **164**, 506–14 (1969). A description of the major approaches taken to combat viruses. Includes detailed discussion of mode of interferon action.
3. ISAACS, A. and LINDENMANN, J., Virus interference. I. The interferon. *Proc. Roy. Soc.*, Ser. B, **147**, 258–67 (1957). The discovery of interferon.
4. MARCUS, P. and SALB, J., Molecular basis of interferon action. Inhibition of viral RNA translation. *Virology*, **30**, 502–16 (1966). Research and theory indicating possible connection between interferon and the genetic apparatus of the cell.
5. PARK, J. and BARON, S., Herpetic keratoconjunctivitis: therapy with synthetic double-stranded RNA. *Science*, **162**, 811–13 (1968). Practical application of theory. Synthetic double-stranded RNA caused therapy of viral eye disease.

6. STEIN, A. and LOEBENSTEIN, G., Induction of resistance to tobacco mosaic virus by Poly I: Poly C in plants. *Nature,* **226,** 363–4 (1970). Induction of immunity to viral infection in higher plants.
7. STERN, R., A nuclease from animal serum which hydrolyses double-stranded RNA. *Biochem. Biophys. Res. Commun.,* **41,** 608–14 (1970). Discovery of an enzyme which promises to be a useful tool for the study of interferon.

PART TWO

When attempting to explain how hormones may exert their effects one specific pathway cannot be implicated alone, and as seen in Fig. 31 several possibilities exist and more often than not one hormone has more than one site or mode of action.

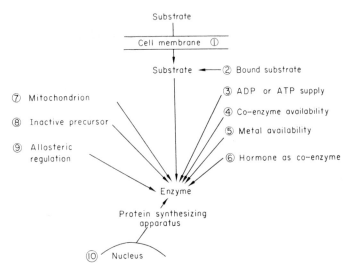

FIG. 31. Possible modes of hormonal action. (After Tepperman[15].)

1. *Effects on membrane systems.* Certain hormones act by enabling substrates or salts to traverse cell membranes or may affect the passage of liquids through membranes. As we shall see in the next chapter, one of the theories pertaining to effects of auxin is in this context. The mammalian hormone insulin increases the permeability of the cell membrane to sugars and the hormone vasopressin in the kidneys affects the passage of Na^+ and K^+.

2. Most of the other possibilities are centred around enzyme production or action and one thus presumes that hormones may be instrumental in release of bound substrate for a pre-existing enzyme, thus triggering enzymatic action.

3, 4, 5. The possibility of hormonal effect by supplying limiting factors which could set the pace of enzymatic reactions. This includes the hormonal control of ATP or ADP whose energy-rich phosphate bonds are a prime source of cell energy. Hormones may also increase availability of metal ions which play a central role in action of many enzymes, e.g. it is known that the hormone produced by the thyroid gland, thyroxin, acts by increasing availability of certain metallic ions and in this respect acts as a chelating agent. Hormones may also affect availability of several co-enzymes.

6. The hormone itself may be a co-enzyme. It was once believed that estrogens could serve as co-enzymes. This is no longer of wide acceptance but the "steroid recognizing proteins" have drawn much attention and the activity of the steroid-protein conjugate in some way resembles the enzyme–coenzyme complex. In the case of the major phytohormones —auxin, gibberellin and kinetin—reports have been made of hormone–protein association essential for growth regulation. The nature of this association will be dealt with in the chapters on each hormone.

7. It is possible that hormonal action might take place in the mitochondrion and the major overall effect may be on respiration. In the event of interaction with mitochondrial components the mode of hormonal effect may be via any of the mechanisms mentioned above.

8. This possibility takes into account the fact that certain enzymes may exist in a biologically inactive state or as a precursor which as yet has still to undergo further changes before the final active form is attained. Usually a state of dynamic equilibrium is reached between the precursor and the final product and it is known that hormones may affect this equilibrium. This category of hormonal effects includes the now well-documented activation of phosphorylase by the hormone epinephrine in muscle tissue. This activation is not direct and the mediator has been found to be the nucleotide 3′5′-*cyclic adenosine monophosphate* (cAMP), the structure of which is shown in Fig. 32. The inactive precursor in this case was found to be diphospho-phosphorylase and now several other hormones besides the epinephrine have been found to act in a like manner. Research along these lines has also

extended to the plant kingdom and this shall be detailed in the chapter dealing with the gibberellins.

Sutherland and Rall,[14] based on this and other observations, have suggested a new concept of hormone action which they term *deputy molecule action*. This envisages the hormone as a *first messenger* (not to be confused with mRNA) which acts at some specific site—often membrane associated—in the cell and there activates another molecule. The latter is considered to be the enzyme *adenyl cyclase* which acts on ATP and causes the production of cAMP. The cAMP thus formed is termed the *second messenger* or the *deputy* of the hormone, and may now activate one or several other proteins present in the cell. These proteins include enzymes, enzyme precursors or CAP protein (see p. 24).

Fɪɢ. 32. 3′5′-Cyclic adenosine monophosphate.

Butcher *et al.*[4] have now suggested an overall scheme which in brief may be summarized as follows. After release from the site of production, a hormone may be transported to its site of action where it interacts with the *adenyl cyclase* system. Its action causes a *membrane-bound* enzyme catalysed cyclization of ATP to cyclic AMP and the release of pyrophosphate. In many but not all tissues of higher species adenyl cyclase has been found associated with membrane fractions. A second enzyme, *cyclic nucleotide phosphodiesterase*, inactivates cAMP by converting it to ordinary 5′ AMP. Recent evidence has indicated that cytokinin-directed plant cell division may be achieved through inhibition of this enzyme. This will be discussed in detail under the heading "Cytokinesins" in Chapter 14.

Butcher *et al.* postulate that a possible model of the adenyl cyclase

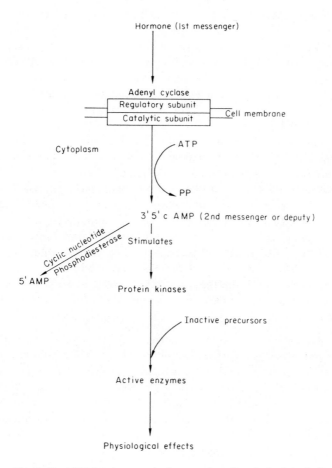

FIG. 33. Cyclic AMP involvement in hormonal mediated physiological responses.

system which at least provides a useful working hypothesis is that it consists of at least two types of subunits: a regulating subunit facing the extracellular fluid and a catalytic subunit with its active centre bordering on the interior of the cell. Interaction between part of the regulatory subunit and a relevant hormone leads to a conformational perturbation which is extended to the catalytic subunit, altering the activity of the latter.

Several research groups have indicated that *protein kinases* are stimulated by cAMP. These kinases are present in many tissues and catalyse phosphorylation of enzyme precursors, casein, prolamine and histones and this may be a unifying concept in the molecular mechanism of cyclic nucleotide action. The above processes are schematically presented in Fig. 33.

A somewhat different approach, but not necessarily different to that outlined in Fig. 33, is the possible interaction with the protein-synthesizing apparatus. This has been discussed in Chapter 4 and the reader is referred to the scheme of Pastan and Perlman presented in Fig. 13 on p. 30. The participation of nuclear components in cAMP-induced effects was also indicated by use of specific antibiotics; for example, Jost *et al.*[7] reported that in addition to the promotion of activity caused by cAMP a marked increase in their total amounts was also observed. This increase could be effectively inhibited by application of actinomycin D. From these and other investigations it is concluded that cAMP action may also be at a transcriptional or a translational level.

9. *Allosteric regulation.* This approach to hormonal regulation attributes properties of allosteric effectors (see Fig. 14) to certain hormones. These may interact with biologically active proteins such as enzymes or molecular repressors mentioned in Chapter 4. Hormones in this case may either be positive or negative effectors. As will later be outlined in the sections dealing with the individual plant hormones it is noted that auxin is considered by some to be a positive allosteric effector and abscisic acid a negative one.

10. *Hormonal effect on the enzyme synthesizing apparatus.* During recent years a great deal of research has been devoted to various aspects of hormonal regulation at a molecular level. It is clear that many possibilities exist and the process as a whole is open to hormonal regulation at many sites. These include the DNA itself or any of the factors participating in transcription or translation, and even if it is known that hormonal regulation is at the level of protein synthesis it is not always possible to state exactly where this takes place.

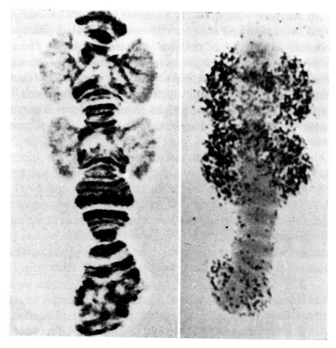

PLATE 7

Puffing of chromosome 4 of the salivary gland of the gnat *Chironomus tentans*.
Left: The chromosome with three puffed regions. These regions are called
Balbiani rings. *Right:* Autoradiograph of the puffed chromosome indicating
that uridine C^{14}, which is specific to RNA, is mainly in the puffed regions,
while non-puffed regions incorporated almost no label. (Reproduced with the
kind permission of Prof. W. Beerman, Max Planck Institute, Tübingen,
Germany.)

It is possible that DNA replication is effected via hormonal regulation
of DNA polymerase (I or II!) and it is also likely that the process is
controlled by the hormonal control of "molecular derepressors". The
derepression of DNA could also result in increased transcription, synthesis of new RNA, etc. When discussing this subject the "histone
hypothesis" of transcriptional control, mentioned on pp. 33–34, should
be kept in mind.

A typical case of direct activation of the chromosome by a hormone

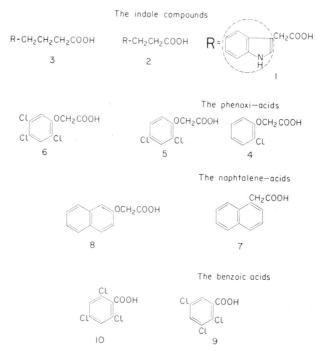

FIG. 35. The main groups of compounds with auxin-like activity. 1. Indole acetic acid (IAA). 2. Indole pyruvic acid (IPA). 3. Indole butyric acid (IBA). 4. Chlorophenoxyacetic acid. 5. 2,4,-Dichlorophenoxy acetic acid (2,4,D). 6. 2,4,5-Trichlorophenoxy acetic acid (2,4,5,T). 7. Naphthaleneacetic acid (NAA). 8. β-Naphthoxyacetic acid. 9. 2,3,6-Trichlorobenzoic acid. 10. 2,4,6-Trichlorobenzoic acid.

(a) *Transcriptional regulation by membrane-bound IAA acceptor protein.* The process of transcription as outlined in Chapter 4 and especially the RNA polymerase "transcription factors" mentioned on page 24 provide the background for much recent research dealing with the possible mode of auxin action. Matthyse and Phillips[13] have reported that in isolated plant nuclei and in isolated chromatin RNA synthesis is promoted by an auxin protein complex. The increased rate of RNA synthesis occurred even in the presence of saturating amounts of RNA polymerase; however, the hormone protein complex did not affect the possibility that hormonal control of protein synthesis may therefore be via regulation of polysome content in the cell.

(d) *Effects on enzyme systems associated with cell walls.* Elucidation of primary auxin action has extended to realms beyond the nucleus and many have sought the solution in one or more enzymatic or structural components of the cell wall. Trewavas[23] has reviewed this subject and in general it may be said that auxin action may intrinsically be connected with some of the above-mentioned processes.

Concerning the nature of cross-linking believed to be responsible for the stability of cell walls, in addition to the usual cellulose, hemicelluloses and pectinic compounds and binding mentioned in standard texts on plant physiology, Lamport[10] has demonstrated the existence of a glycoprotein, termed *extensin*, which is potentially capable of cross-linking a hugely disproportionate amount of wall polysaccharide. The linkage has been found to be O-glycosidic bonding between the amino acid *hydroxyproline* and arabinose. In cultured cells of tomatoes, for example, 90 per cent or more of the hydroxyproline is glycosylated and it appears that as a rule the level of hydroxyproline reflects the extent of cross-linking. It is also of interest to note that young extensible tissues have a comparatively low hydroxyproline content while in older tissues the content is high. In the following paragraphs we shall discuss the underlying principles upon which the approach that auxin action is on cell-wall enzymes is based.

When considering the auxin-induced elongation of cereal coleoptiles, one of the most extensive bioassays employed to study the hormone's action, it is observed that the process proceeds in two stages:

(i) Increase of plasticity of cell walls. Plasticity is the ability of tissues to extend irreversibly as opposed to elasticity which is reversible. This increase is apparently due to breakage of cross-links between the cellulose microfibrils in the wall, a step requiring both auxin and oxygen.

The links broken may be the hydroxyproline–arabinose ones mentioned above and what essentially happens during this stage has been likened by R. Cleland to the process of vulcanization of crude rubber, where not only is linkage broken but also reformed in such a way allowing greater extensibility. This is indicated schematically as follows:

before processing   after

(ii) Osmotic uptake of water causing an increase of cell volume and hence growth. This step requires neither auxin nor oxygen.

This being so, many foremost workers in the plant hormone field have concerned themselves with stage (i), i.e. the increase of cell-wall plasticity and have relegated the initial auxin effect to some plasticity-connected mechanism. A central question posed here is whether the "softening-up" of the cell wall is due to enzymatic digestion, physical changes such as surface tension or some other process. Before an attempt is made to answer it is essential to briefly explain the meaning of a *lag period* in hormonal action. The lag is considered the minimal period of time elapsing between hormonal application to a tissue and its detectable physiological response. In the present discussion this is the period between auxin application to coleoptiles and the commencement of elongation. Since the termination of the lag is indicated by an induced response it stands to reason that any effect, molecular or other, of the hormone on basic plant metabolism must have occurred during this period. In the auxin-coleoptile system the lag is usually of the order of several minutes but under conditions of high auxin concentrations and elevated temperatures there is essentially a zero time lag.[15]

Assuming the very brief lag under normal conditions it is clear that processes leading to cell elongation must have been consummated within a matter of minutes and much detailed research pertains to observed changes during the lag. Masuda and Kamisaka,[12] for example, contend that auxin induces the appearance of a specific mRNA species during the lag. This mRNA finally expresses itself as enzyme participating in cell wall hydrolysis, one of the candidates for which is $\beta 1,3$-*glucanase* which acts on hemicelluloses, loosens the walls and increases plasticity. This contention is in accordance with the "two-staged" elongation outlined above and is essentially associated with the initial step. Masuda and Kamisaka have presented a model in which auxin primarily acts on "primer" DNA producing the specific mRNA which codes for enzymes such as glucanase. The suggested steps may be summarized as follows:

$$\underset{\text{"primer"}}{\text{DNA}} \xrightarrow[\text{auxin}]{\text{activated by}} \text{specific mRNA} \rightarrow \beta1,3\text{-glucanase} \xrightarrow[\text{loosening}]{\text{cell wall}} \rightarrow \text{growth.}$$

While this scheme makes things somewhat clearer no explanation is offered as to how the auxin activates the DNA. The reader may assume that this could be via one of the mechanisms mentioned by Fellenberg (p. 110) and is reminded of the observation reported in Chapter 4 that IAA in *Lolium* is associated with arginine-rich histones which are considered more inhibitory than the lysine-rich ones.

The intervention of cell-wall enzymes in auxin-induction elongation has been challenged by the observations that while the lag may normally be of the order of 10 minutes or less, detectable endogenous changes in enzyme content only appear much later. This implies that primary auxin action has no bearing on cell-wall enzymes and that these may appear at a later stage in the elongation process. This claim does not conclusively disprove the point since exogenously applied glucanase induced elongation after a lag period only slightly longer than the normal one and, furthermore, endogenous concentrations of the enzyme-promoting elongation may be at a level too low for detection by standard biochemical procedures.

While the above discussion has mainly concerned itself with glucanase it is pointed out that this is only one of a series of enzymes which have been suggested as major contributing factors in auxin-induced cell elongation. Other such enzymes include *cellulase, pectinase* and *pectin methyl-esterase*, the latter two possibly acting on the pectic substances in the middle lamella which "cements" two adjacent cells together. The pectin methyl-esterase could possibly esterify the —COO—Ca—OOC— bonding obtained by linkage of free carboxylic groups of galacturonic acids by divalent metal ions such as Ca. This bond upon being severed by esterification could increase plasticity.

Kinetic studies on the coleoptile-elongation system carried out by Zenk and Nissel[15] and Evans and Ray[5] under specific conditions utilizing delicate measuring procedures indicated no measurable "lag" period since response to auxin was immediate. These workers have therefore reached the conclusion that auxin promotes neither transcription nor translation, has no effect on protein synthesis and its physiological effect is not related to nucleic-acid metabolism. These conclusions were based on detailed studies of the dependence of lag on auxin concentration, temperature and the "steady state" of elongation as a function of auxin content. Theoretical mathematical calculations, beyond the scope of the present text, were employed to prove that under the experimental conditions used, no possibility existed that auxin action was via gene activation. In their opinion all reported observations of the appearance of new or more nucleic acids, or binding with such, are later stages of auxin action occurring when cells have already begun to multiply. The general conclusion drawn is that the site of auxin action is on some pre-existing cellular system and one possibility suggested by them is allosteric activation of one of the enzymes participating in hydrolysis or synthesis of the polymers comprising the cell wall.

According to this approach elongation is a series of stages connected with the loosening of cell walls, growth of cells by osmotic uptake of water and subsequent repair of the sundered cell-wall links. Besides the suggested activation of pre-existing enzymes it is also postulated that auxin may have some direct effect on the permeability of the cell membrane, thus enhancing water uptake.

Pope and Black[18] have also suggested that extension growth produced by auxin may be by means of incorporation of a structural factor, possibly protein from a *previously* synthesized pool into cell walls. They also produced evidence against the concept that auxin promotes cell extension by inducing protein synthesis. Their conclusions were based on the observation that the antibiotic cyclohexamide inhibited over 90 per cent ^{14}C leucine incorporation into protein within 10 minutes after application and that IAA induces growth in the presence of the inhibitor.

(e) *Auxin and the Golgi apparatus.* Electron microscopic studies carried out by Frey-Wyssling have demonstrated that auxin induces an increased rate of secretion of vesicles from the Golgi apparatus in the cytoplasm (Plate 8). These vesicles contain carbohydrate and usually migrate from the Golgi to the plasmalemma where their contents provide the matrix in which the cell wall is formed. This observation is a further indication that auxin action is connected with cell-wall metabolism and suggests that the site of initial action is not the wall itself.

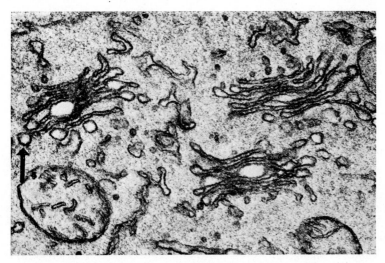

PLATE 8
Electron micrograph of castor bean *Ricinus communis* cells indicating Golgi apparata (folded membranes) secreting vesicles containing carbohydrate and migrating to cell periphery where they form the matrix of the cell wall. According to Frey-Wyssling one of the effects of auxin is the enhanced rate of vesicle secretion and migration. Arrow indicates a typical vesicle. Enlargement × 40,000. (This plate is reproduced with the kind permission of Prof. Frey-Wyssling, Laboratory of Electron Microscopy, Advanced Technical Institute, Zürich, Switzerland.)

(f) *Allosteric effect of auxin.* The opinion that auxin may act as an allosteric effector (see p. 32) is a comparatively new approach to the action of this phytohormone and conclusive experimental demonstration that this is indeed the case lags far behind the theoretical models constructed about it. Nevertheless, some rather interesting leads have been provided and the possibility of allosteric hormonal regulation both positive and negative has been extended to other hormones as well. Sarkissian,[20] who made one of the earliest reports on the subject, conducted his experiments on the enzyme citrate synthase which was isolated from the scuttela of corn grains. This enzyme catalyses the conversion of acetyl co-enzyme A to citric acid at the metabolic "junction point" of glycolysis and the Krebs cycle during respiration, and was chosen as an object for study since it is considered to be a "regula-

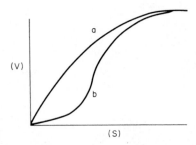

FIG. 38. Activity curves for normal (a) and "regulatory" (b) enzymes.

tory" enzyme. This means that in addition to its site of attachment to its substrate, which in the present case is acetyl co-enzyme A, it also possesses a site for attachment of an allosteric effector (see Fig. 14 in Chapter 4). In a graph in which one arm represents substrate concentration (S) and the other the velocity of the reaction (V), citrate synthase exhibits a sigmoidal behaviour which is considered typical of regulatory enzymes and in this respect differs essentially from the normal curve obtained with enzymes lacking allosteric sites (see Fig. 38). By addition of IAA to the enzyme Sarkissian indicated an increase in the latter's activity and also an accentuation of the sigmoidal shape of the curve is obtained (see Fig. 39). These phenomena he believes are an indication of a positive allosteric effect of the hormone. As explained in Part One of this book, the allosteric sensitivity of active proteins can be nullified by certain means and its loss does not necessarily impair their ability to attach to substrates. By freezing and thawing the fresh enzymes Sarkissian attempted to achieve this, and reported that the frozen-and-thawed enzyme did not respond to IAA as did the fresh preparation. Upon investigating possible allosteric sites on the enzyme it was furthermore claimed that these were associated with SH groups which confer stability upon the secondary structure of the enzyme protein. This conclusion was drawn upon comparing the nature of the sulphydryl binding of the fresh and 8-hour frozen and subsequently thawed enzyme which had lost its allosteric sensitivity. As seen in Table 5 the relative prevalence of —SH— and —SS— groups was determined and it appears that loss of allosteric sensitivity is accompanied by disappearance of the

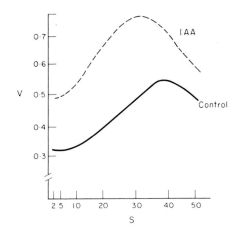

Fig. 39. Activation of citrate synthase activity by IAA as a function of substrate concentration. IAA concentration was 1.25×10^{-11} M. Substrate concentration (S) expressed as μM acetyl coenzyme A $\times 10^{-3}$. Velocity of reaction (V) presented as relative units. (Data of Sarkissian[20].)

sulphydryl (SH) groups which apparently are converted to disulphide (SS) ones.

The findings are in Sarkissian's opinion the solution to the primary action of auxin in plants. It is not claimed that citrate synthase is the specific and key enzyme affected by the hormone, but rather it is representative of a whole series of sulphydryl regulatory enzymes of wide occurrence in plant tissues. The phenomena reported in sections (a)–(d)

TABLE 5. QUANTITATIVE COMPARISON OF SULPHYDRYL (SH) AND DISULPHIDE (SS) GROUPS IN FRESH AND FROZEN-AND-THAWED CITRATE SYNTHASE

(According to Sarkissian[20])

Preparation tested	Nanomols/mg protein	
	SH	SS
Fresh enzyme	7·5	22·5
Enzyme frozen for 8 hours and thawed	0·0	30·0

in this chapter are not ignored but considered as secondary effects caused by the initial allosteric activation by the auxin of its endogenous acceptors—SH groups in biologically active proteins.

RECOMMENDED READING AND REFERENCES

1. BENDANA, F. E., GALSTON, A. W., SAWNHEY, R. K. and PENNY, P. J., Recovery of labelled ribonucleic acid following administration of labelled auxin to green pea stems. *Plant Phys.* **406**, 977–83 (1965). One of the foundations upon which the "IAA–RNA complex" concept rests. An interesting discussion of how such a complex could influence basic cell metabolism.
2. DATTA, B. and BISWAS, A., Effect of indoleacetic acid on protein and ribonucleic acid synthesis. *Experientia* **21**, 633–4 (1965). A demonstration that tRNA is a likely candidate for IAA complexing.
3. DAVIES, P. J., Evidence against an IAA–tRNA complex. *Plant Cell Phys.* **12**, 785–9 (1971). See no. 4 following.
4. DAVIES, P. J. and GALSTON, A. W., Labelled indole macromolecular conjugates from growing stems supplied with labelled indoleacetic acid. I. Fractionation. *Plant Phys.* **47**, 435–41 (1971). A re-evaluation of IAA–RNA complex suggesting that the macromolecule bound to auxin is not nucleic acid.
5. EVANS, M. L. and RAY, P. M., Timing of the auxin response in coleoptiles and its implications regarding auxin action. *J. Gen. Phys.* **53**, 1–20 (1969). A series of experiments leading to the conclusion that primary action of auxin is not upon the genetic apparatus and inasmuch as this is encountered it is a late and secondary effect.
6. FELLENBERG, G., Veränderungen des Nucleoproteids unter dem Einfluss von Auxin und Ascorbinsäure bei der Wurzelneubildung an Erbsenepikotylen. *Planta* **84**, 324–38 (1969). Auxin may act as a nucleic derepressor possibly by effecting histones.
7. GAYLER, K. R. and GLASZIOU, K. T., Plant enzyme synthesis. Hormonal regulation of invertase and peroxidase synthesis in sugar cane. *Planta* **84**, 185–94 (1969). mRNA may be stabilized by auxin.
8. HARDIN, J., CHERRY, J., MORRE, D. and LEMBI, C., Enhancement of RNA polymerase activity by a factor released by auxin from plasma membrane. *Proc. Nat. Acad. Sci.* **69**, 3146–50 (1972). A basic contribution indicating connection between auxin action, the plasmalemma and the process of transcription.
9. HERTEL, R., THOMPSON, K. S. and RUSSO, V., *In vitro* auxin binding to particulate cell fractions from corn coleoptiles. *Planta* **107**, 325–40 (1972). Indirect evidence that the auxin acceptor may be a membrane-associated cell particle.
10. LAMPORT, D., Hydroxyproline-O-glycosidic linkage of the plant cell wall glycoprotein extension. *Nature* **216**, 1322–4 (1967). The discovery of a wall protein playing a central role in plasticity-connected phenomena.
11. MASUDA, Y., Role of cellular ribonucleic acid in the growth response of *Avena* coleoptile to auxin. *Phys. Pl.* **12**, 324–35 (1959). A pioneering effort towards the relegation of hormonal action to processes involving nucleic acid metabolism.
12. MASUDA, Y. and KAMISAKA, S., Rapid stimulation of RNA biosynthesis by auxin. *Plant Cell Phys.* **10**, 79–86 (1969). A typical research pointing to the fact that auxin acts on cell wall loosening enzymes. The possibility that this may be via activation of nucleic acids is not overruled.

13. MATTHYSE, A. and PHILLIPS, C., A protein intermediary in the interaction of a hormone with the genome. *Proc. Nat. Acad. Sci.* **63**, 897–903 (1969). An interesting observation at a molecular level which shows that an auxin:protein complex makes an increased portion of the genome available for transcription.

14. MONDAL, H., MANDAL, R. and BISWAS, B., RNA stimulated by indole acetic acid. *Nature, New. Biol.* **240**, 111–13 (1972). Together with no. 8 above a clear indication that at least one basic mode of auxin action is linked to the process of transcription, probably at the initiation stage.

15. NISSEL, D. and ZENK, M., Evidence against induction of protein synthesis during auxin induced initial elongation of Avena coleoptiles. *Planta* **89**, 323–41 (1969). See no. 5 above.

16. PENNY, P. and GALSTON, A. W., The kinetics of inhibition of auxin induced growth in green pea segments by actinomycin D and other substances. *Am. J. Bot.* **53**, 1–7 (1965). A pioneering effort indicating interactions between phytohormones and nucleic acids.

17. PILET, P. E. and BRAUN, R., Ribonuclease activity and auxin effects in the *Lens* root. *Phys. Pl.* **23**, 245–50 (1970). IAA may control RNA levels in plants by regulation of RNase.

18. POPE, D. and BLACK, M., The effect of indole-3-acetic acid on coleoptile extension growth in the absence of protein synthesis. *Planta* **102**, 26–36 (1972). Indirect evidence that protein synthesis may not be involved in auxin produced growth promotion.

19. ROYCHOUDRY, R. and CHEN, S. P., Studies on the mechanism of auxin action: auxin regulation of nucleic acid metabolism in pea internodes and coconut milk nuclei. *Phys. Plant.* **17**, 352–62 (1964). The possibility of IAA–DNA binding discussed.

20. SARKISSIAN, I. V., Nature of molecular action of 3-indole acetic acid, in *Biochemistry and Physiology of Plant Growth Substances*, pp. 473–85, ed. P. WIGHTMAN and G. SETTERFIELD, Runge Press, Ottawa (1968). A most pertinent presentation of the controversial hypothesis of auxin as an allosteric effector.

21. THOMPSON, W. F. and CLELAND, R., Auxin and ribonucleic acid synthesis in pea stem tissue as studied by deoxyribonucleic acid-ribonucleic acid hybridization. *Pl. Phys.* **48**, 663–70 (1971). Utilization of the hybridization technique to provide evidence in support of the view that auxin does not necessarily exert its effect by induction of production of a new RNA species.

22. TREWAVAS, A. J. T., Effect of IAA on RNA and protein synthesis. *Arch. Biochem. Biophys.* **123**, 324–35 (1968). Further candidates for auxin promotion—the polysomes.

23. TREWAVAS, A., Relationship between plant growth hormones and nucleic acid metabolism. See Vol. 2, pp. 161–92 in reference no. 12 of Chapter 12.

24. VAN DER WOUDE, W., LEMBI, C. and MORRE, D., Auxin (2,4-d) stimulation (*in vivo* and *in vitro*) of polysaccharide synthesis in plasma membrane fragments isolated from onion stems. *Biochem. Biophys. Res. Comm.* **46**, 245–53 (1972). Evidence for membrane-associated auxin action.

25. VENIS, M., Promotion of deoxyribonucleic acid-dependent ribonucleic acid synthesis by protein isolated on a plant hormone affinity column. *Biochem. J.* **127**, 29. A short note reporting transcriptional promotion by a protein released from an auxin complex.

26. YAMAKI, T. and KOBAYASHI, K., Radioactive indoleacetic acid in sRNA fraction, in *Plant Growth Substances*, 1970. *Proc. Int. Cong. in Growth Substances*, Canberra. Abstracts. Recent evidence that IAA may indeed complex with tRNA.

CHAPTER 14

The Mode of Cytokinin Action

THE cytokinins are a group of growth substances which are usually derivatives of the nucleic purine base adenine. However, other substances such as diphenyl urea or derivatives of nicotinamide are also able to promote the process of cell division or *cytokinesis* (Wood *et al.*[14]). Another distinct generic group of cell-promoting factors termed *cytokinesins* will be discussed separately at the end of this chapter. The structures of some of the purine derivatives possessing cytokinin activity are presented in Fig. 40. Kinetin and benzyladenine are synthetic substances while zeatin and isopentenyladenine (IPA) are endogenous cytokinins found in higher plants, yeasts and bacteria.

Since adenine is a component of this group of compounds their mode of action was sought for in the field of nucleic acid metabolism. The possibility that cytokinin action is connected with RNA was based upon the determination of Zachau *et al.*[16] that the natural cytokinin IPA is present in tRNA specific for the amino acid serine and within the adaptor molecule is located adjacent to the anticodon (see Fig. 16). It has since been found that IPA occurs in the tRNA of a wide variety of organisms including bacteria such as *E. coli* and *Cornebacterium fasciens*, yeasts, higher plants and mammals. In Part One it was pointed out that for each amino acid there exists at least one specific type of tRNA and it now has been demonstrated that IPA is not present in all the types. To date its presence has been reported in tRNA's for serine, isoleucine and tyrosine while it is absent from tRNA's for arginine, glycine, phenylalanine and valine.

Skoog *et al.*[12] found that tRNA derived from yeast, liver or from *E. coli* was able to promote growth of tobacco callus tissue growing in culture, this being considered a specific cytokinin promoted phenomenon. No growth response was elicited by application of rRNA, and

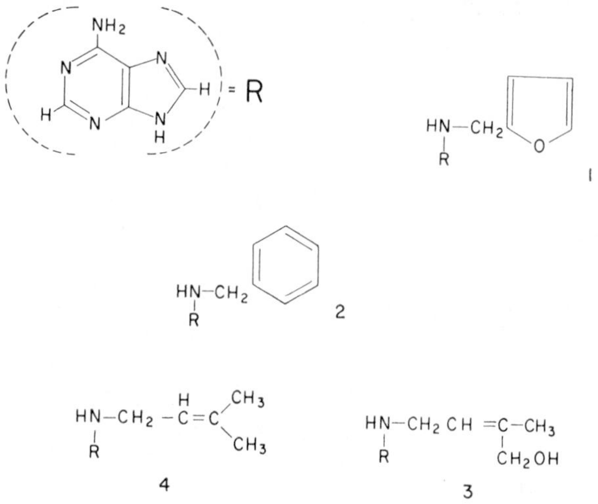

FIG. 40. Substances with cytokinin activity. 1. Kinetin or furfuryl amino-purine. 2. Benzyl-adenine or 6-benzyl aminopurine (BAP). 3. Zeatin or 6-(4-hydroxy-3-methyl-trans-2-anyl) aminopurine. 4. Isopentenyladenine (IPA) or 6-(yy-dimethylallyl) aminopurine.

maximal promotion was traced to serine tRNA, i.e. an IPA containing molecule. As would be expected, no callus-growth promotion was obtained from tRNA's for arginine, glycine, phenylalanine or valine, all of which lack the IPA. Skoog *et al.* state that one molecule of IPA per twenty molecules tRNA is sufficient to cause growth responses as experienced in callus culture, and thus in general the number of molecules per cell responsible for cytokinin activity is extremely small.

Pursuing the matter further Fittler and Hall[1] state that the presence of IPA adjacent to the anti-codon is necessary for the latter's recognition of the codon on the mRNA. This conclusion was reached since iodine, which combines specifically with IPA and prevents the binding of aminoacyl tRNA to mRNA, did not interfere with the loading of the specific amino acid onto the tRNA.

This approach leads to the logical question: Is the presence of the cytokinin in tRNA responsible for its hormonal activity or is it only a structural component of the tRNA and hormonal action dependent upon other processes? Evidence supporting both points of view has been produced and in brief we shall discuss the salient features of each commencing with data supporting the former hypothesis.

As stated previously, callus tissue, which is an undifferentiated mass of cells which forms on wound surfaces of certain plants, requires cytokinin in order to develop. If the synthetic cytokinin benzyladenine is added to callus tissue in culture most of the added growth substance is broken down but about 15 per cent remains intact and can be detected in

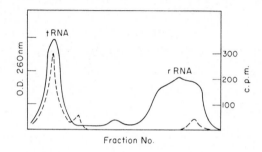

FIG. 41. The incorporation of labelled BAP into tRNA of soybean callus culture. Continuous lines, RNA profile as determined by MAK fractionation. Broken lines, radioactivity. (After Fox[2].)

tRNA (see Fig. 41). Moreover, Fox and Chen[4] claim that IPA performs a role similar to that of an RNA-methylating enzyme and that the incorporation of IPA into tRNA, like its methylation, facilitates the unloading of amino acids from the adaptor molecule. In this claim Fox and Chen claim support from the hypothesis stating that the lack of methylated bases in tRNA causes incorrect reading of the codon. The substitution of methylation by cytokinin at present remains an interesting but experimentally unproved hypothesis. In this context it is of passing interest to mention that in ageing nematode tissues it has been demonstrated (Gershon and Gershon[6]) that there is an accumulation of non-functional enzyme molecules. This indicates that in ageing tissues correct codons may be misread by tRNA during translation.

The alternative approach that hormonal action of cytokinin is not dependent upon its presence in tRNA is presented typically by Kende and Tavares.[7] They report that while IPA *is* present in tRNA molecules of the bacterium *E. coli*, cytokinin is not necessary for its growth and they consider it very unlikely that cytokinin serves as a precursor of the bacterial IPA. Employing computer-aided mathematical calculations they claimed that the probability of obtaining an *E. coli* mutation which would have an obligate ("auxotrophic") cytokinin requirement is practically zero—one in 10^{105}. They propose that IPA is not formed from cytokinin but that initially adenine exists as an integral part of the tRNA molecule and to this an isopentenyl side chain is added. This side chain is formed from compounds unconnected with cytokinin, e.g. mevalonic acid or isopentenyl pyrophosphate. It is also known that when tRNA is broken down nucleotides are released and it is possible that cytokinins are set free in this manner, migrate to other cellular sites and

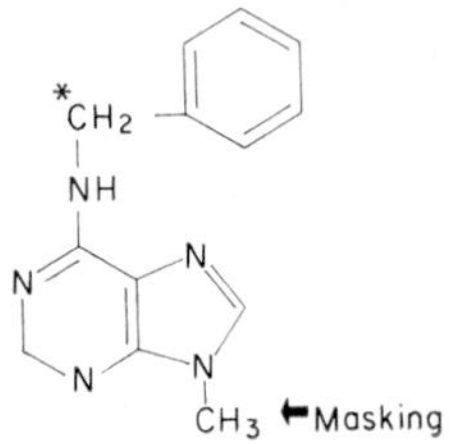

Fig. 42. 6-Benzylamino 9-methyl purine. * = labelled carbon atom.

there function as growth regulators. This physiological effect is experienced only after the cytokinin is freed from tRNA and in a bound form would be inactive. Kende and Tavares performed an experiment in which they attempted to distinguish between the hormonal effect and incorporation into tRNA by application of a "masking" technique. They synthesized a labelled cytokinin, 6-benzylamino 9-*methyl* purine (see Fig. 42), which resembles benzyladenine but differs from it in that it possesses a methyl group on the adenine at serial position 9. This methylation or "masking" of the 9-position prevents the incorporation of the whole molecule into the tRNA of soya callus tissue while as previously seen in Fig. 41 ordinary unmasked benzyladenine is taken up by the tRNA. The failure of the masked cytokinin to be incorporated did not impair its physiological action and the treated callus evidenced marked growth typical of cytokinin. From this experiment it is apparent that lack of association of cytokinin with tRNA does not curtail biological activity and from this and the previously mentioned data no hormonal significance is attributed to the presence of IPA adjacent to the anticodon in the tRNA.

This evidence now appears less conclusive than previously believed since Fox *et al.*[5] have demonstrated that cytokinin requiring tissues *remove the 9-substitution* in as short a time as 10 minutes after application and among the metabolic products free benzyl amino purine, an active cytokinin, appears.

Cytokinins and senescence

A process known to be under the influence of cytokinin is the ageing or senescence of leaves and it has been found that this hormone is one of the factors which may cause senescence retardation. Various explanations, some including molecular processes, have been offered to account for the phenomenon. It has been established that in senescing tissues levels of DNA, RNA and protein gradually decrease and that RNA: DNA ratio changes. It has been found (see review by Fox[3]) that cytokinin retards these processes and also stabilizes or even promotes RNA:DNA ratios, thus indicating that RNA metabolism may be directly involved. Concerning the specific species of RNA affected by cytokinin during senescence Srivastava[13] has suggested that ageing is not a passive but rather an active process during which specific RNA molecules coding for catabolic enzymes participating in senescence are produced. These include RNase, peptidases and cellulases, all of which contribute to cell degradation.

This suggestion may be disputed on the grounds that the above newly formed enzymes are proteins and thus an *increase* of protein content should result and not the reported *decrease*. Srivastava's answer to this contention is an observation made on senescing barley leaves that during the initial few days leaf protein content actually increases and only at a later stage does the content fall. He believes that during the initial period the degradative proteins are formed and that they subsequently cause the decrease noticed during the later stages.

According to the above hypothesis, cytokinin acts by preventing the synthesis of specific mRNA coding for degradative enzymes. A somewhat slightly different interpretation of RNA involvement in senescence is given by Phillips *et al.*[9] who also consider the role of degradative enzymes—chlorophyllase and RNase—in the process and it appears that high levels of these enzymes may represent high turnover rates, and not necessarily synthesis of new mRNA coding for them.

Richmond *et al.*[10] have attempted here too to ascertain whether cytokinin's action in senescence depends on its structural site in the tRNA and have performed a "masking" experiment similar in rationale to that of Kende and Tavares.[7] The cytokinin 6-benzyl amino purine benzyl $C^{14}7$ which was not incorporated into tRNA was highly active in senescence retarding, leading to the conclusion that the hormonal effect is independent. However, here again the possibility exists that the mask may be removed enzymatically and that finally incorporation does occur to a certain degree.

Concerning the overall regulation of plant senescence by cytokinin Sitton *et al.*[11] have presented an interesting theory. It is well documented that cytokinin serves as a metabolic *sink* "attracting" metabolites and certain amino acids and it is also known that root tissues are rich in cytokinins. Senescence may therefore be connected with the decreased cytokinin supply from root to shoot which may be accompanied by appearance of new cytokinin "sources"—the developing seeds in immature fruit. These draw on shoot metabolites and amino acids and so doing promote senescence. As yet no conclusive molecular or other theory of the mechanism of cytokinin-induced "metabolite attraction" has been offered although the phenomenon itself has been reported for a wide variety of plant species.

Cytokinesins

Two members of a newly described class of cell division promoting factors with partition coefficients or K values of 1·9 and 2·7 in butanol–water after 500 transfers in countercurrent distribution have been isolated from crown-gall tumor tissue of *Vinca rosea* and have respectively been designated *cytokinesin I* and *cytokinesin II* (Wood *et al.*[15]). According to these workers cytokinesins should not be considered members of the cytokinin group since the latter are by definition only natural or synthetic N^6-monosubstituted adenyl derivatives. Physical and chemical studies have demonstrated that cytokinins are *purinone* derivatives that contain alkyl groups and sulphur, and glucose as the sugar moiety. They differ therefore from the substituted purine cytokinins isolated from certain tRNA's mentioned earlier in this chapter which contain ribose and not glucose.

Wood and his co-workers suggest that the classic adenylate cytokinins activate the synthesis of the purinone derivatives in normal cell types and it is the purinones, *rather than the cytokinins*, that are directly involved in the promotion of cell division. The suggested mode of action in brief is as follows:

These substances and in particular cytokinesin I are potent inhibitors of the enzyme *cyclic nucleotide phosphodiesterase* (see Fig. 33) of both plants and animals. This observation suggests that cytokinesin may exert its biological effects in promoting division of plant cells via its control of 3'5'-cAMP levels. It is also of interest to note in context with the possible key role of the enzyme cyclic nucleotide phosphodiesterase in the cell division process that *theophylline*, belonging to the methyl-xanthine group known to be inhibitory to phosphodiesterase, promotes cell division in tobacco pith cultures.

These findings indicate that in the case of cytokinins, as well as the other major groups of phytohormones, cAMP metabolism may be of prime importance in manifestation of physiological effects.

RECOMMENDED READING AND REFERENCES

1. FITTLER, F. and HALL, R., Selective modification of yeast seryl-t-RNA and its effect on the acceptance and binding functions. *Biochem. Biophys. Res. Commun.*, **25**, 441–6 (1966). The presentation of the theory that cytokinin action is connected with codon–anticodon recognition.
2. Fox, J., Incorporation of a kinin N,6 benzyl adenine into soluble RNA, *Pl. Physiol.*, **41**, 75–82 (1966). A further argument in support of the hypothesis that cytokinin's hormonal effects depend upon its presence in RNA.
3. Fox, J., The cytokinins. An excellent review article on physiology and biochemistry of the cytokinins on pp. 85–123 of reference no. 17 in Chapter 12.
4. Fox, J. and CHEN, C., Characterization of labelled ribonucleic acid from tissue grown on ^{14}C containing cytokinins. *J. Biol. Chem.*, **242**, 4490–4 (1967). Cytokinin may mimic the action of nucleic acid methylating enzymes.

5. Fox, J., Sood, C., Buckwater, B. and McChesny, J., The metabolism and biological activity of a 9 substituted cytokinin. *Pl. Physiol.*, **47**, 275–81 (1971). "Masking" of a cytokinin does not necessarily prevent its incorporation into nucleic acid as contended in reference no. 6.

6. Gershon, D. and Gershon, H., Detection of inactive molecules in ageing organisms. *Nature*, **227**, 1214–17 (1970). An interesting interpretation of ageing based on faulty information transfer in aged organisms.

7. Kende, H. and Tavares, J. On the significance of cytokinin incorporation into RNA. *Pl. Physiol.*, **43**, 1244–8 (1968). A series of elegant experiments and mathematical calculations suggesting that cytokinin action as a hormone is not connected with its presence in tRNA.

8. Osborne, D., Effect of kinetin on protein and nucleic acid metabolism in *Xanthium* leaves during senescence. *Pl. Physiol.*, **37**, 595–602 (1967). Retardation of senescence is attributed to cytokinin-regulated persistence of RNA synthesis.

9. Phillips, D. R., Horton, R. F. and Fletcher, R. A., Ribonuclease and chlorophyllase activities in senescing leaves. *Phys. Pl.* **22**, 1050–4 (1969). The stress of importance of *turnover* and not only synthesis during senescence.

10. Richmond, A., Back, A. and Sachs, B., A study of the hypothetical role of cytokinins in completion of t-RNA. *Planta*, **90**, 57–65 (1970). Supporting evidence for the independence of cytokinin action, the process investigated being senescence.

11. Sitton, D., Itai, C. and Kende, H., Decreased cytokinin production in the roots as a factor in shoot senescence. *Planta*, **73**, 296–300 (1967). A model for cytokinin control of plant senescence.

12. Skoog, F., Armstrong, D. J., Cherayil, J., Hampel, A. and Bock, R., Cytokinin activity: localization in tRNA preparations. *Science*, **154**, 1354–6 (1966). The story behind the discovery of hormonal action in certain RNA species.

13. Srivastava, B., Mechanism of action of kinetin in the retardation of senescence in excised leaves, pp. 1479–94 (1968). See book referred to in no. 20 in Chapter 13. An interesting research claiming that senescence is an active process involving synthesis of a specific type of RNA.

14. Wood, H. N., Braun, A. C., Brandes, H. and Kende, H., Studies on the distribution and properties of a new class of cell division promoting substances from higher plant species. *Proc. Nat. Acad. Sci.*, **62**, 349–56 (1969). Not only purine derivatives promote cytokinesis. The isolation of a non-purine cytokinin from tobacco, cactus and *Vinca* gall tissue.

15. Wood, H. N., Lin, M. C. and Braun, A. C., The inhibition of plant and animal adenosine 3'5'-cyclic monophosphate phosphodiesterases by a cell division promoting substance from tissues of higher plant species. *Proc. Nat. Acad. Sci.*, **69**, 403–6 (1972). An elegant research and well put argument suggesting that cytokinin-directed plant cell division is mediated by purinone compounds termed cytokinesins.

16. Zachau, H. G., Dutting, D., Feldmann, H., Melchers, F. and Karau, W., Serine specific transfer ribonucleic acids. XIV. Comparison of nucleotide sequences and secondary structure models. *Cold Spring Harbor Symp. Quant. Biol.*, **xxxi**, 417–34 (1966). Biochemical determination of base sequence and secondary structure of tRNA.

CHAPTER 15

Mode of Action of the Gibberellins

THE gibberellins are a large group of substances and at present over forty different types have been characterized. They are all diterpenoids, usually but not always possess five rings and, according to an agreed system of nomenclature, are termed GA_1, GA_2, GA_3, etc. From a structural point of view GA_9 may be regarded as the basic type and all others have the same framework but possess varying degrees of substitution of side groups (see Fig. 43). Gibberellin GA_3, termed gibberellic acid, is one of the prevalent types in higher plants. The figure outlines nine of the series and it should be noted that as in the case of GA_{13} the lactone ring is sometimes missing. It is also believed that some GA's are inactive "discard products" of active forms.

(a) GA and nucleic acid metabolism

Concerning the possible connection of this group of substances and molecular processes probes have been made in various directions. Nitsan and Lang,[11] working on elongation of Lens culinaris (lentil) epicotyls, found a typical GA-induced enhancement of the process. This promotion of elongation was inhibited by the specific DNA antimetabolite 5-fluorodeoxy-uridine (FDUR) and as a rule GA promoted, and FDUR inhibited cell elongation, and synthesis of DNA and RNA. When applied simultaneously GA somewhat reversed the inhibitory effect of the antimetabolite as indicated in Table 6.

In another trial performed by Nitsan and Lang it was demonstrated that the inhibition induced by FDUR could be reversed by the nucleic base thymidine (which as explained in Chapter 2 is specific to DNA and not RNA) but not by application of uridine which is specific to RNA

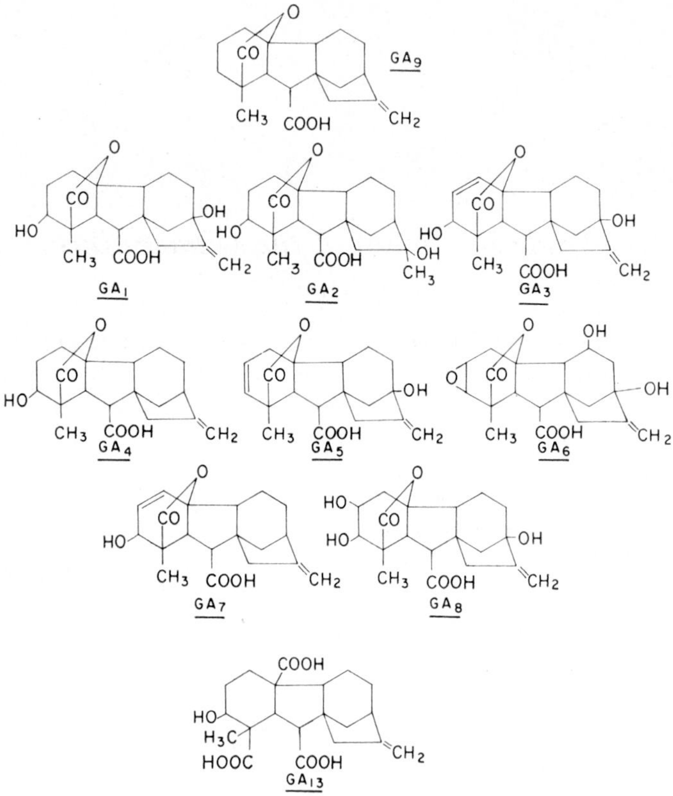

FIG. 43. Some of the various types of gibberellin.

TABLE 6. LENGTH AND CONTENT OF NUCLEIC ACID OF LENTIL EPI-
COTYLS TREATED WITH GA, FDUR OR A COMBINATION OF BOTH
(Data of Nitsan and Lang[11])

Treatment	Epicotyl length (mm)	$\mu\mu$g nucleic acid per cell	
		DNA	RNA
Control	19·0	16·3	69·2
GA 100 ppm	27·6	22·0	90·9
FDUR 10^{-5} M	10·9	12·2	62·1
GA + FDUR	16·9	17·7	71·3

and not DNA. The conclusion drawn from these results is that the GA-induced elongation is dependent upon DNA metabolism.

Continuation of the research along these lines in the same laboratory and by Degani et al.[1] has produced new evidence which suggests that the conclusions of Nitsan and Lang[11] are not as self-evident as their experimental data indicate. Degani et al. found that the reported DNA increment experienced by GA application is mainly of *organelle* and not *nuclear* DNA and could be traced to chloroplasts and mitochondria. Moreover, GA enhancement of incorporation of labelled thymidine was 10–20 times greater in organelle than in nuclear DNA. It has also been reported that upon repeating the lentil epicotyl elongation experiment, growth could be attributed to elongation of the individual cells—a process which does not necessarily entail DNA activity—and was *not* due to cell multiplication which does.

A further challenge to the participation of DNA in GA-induced hormonal responses has been posed by Haber et al.[4] The approach taken by this group is that if the DNA could be inactivated, the response or failure to respond to subsequently applied hormone would indicate whether the genetic material is necessary for physiological response. By irradiating lettuce plantlets with a high dosage of *gamma* rays cell multiplication and DNA synthesis in the tissue was completely shut off. GA applied at this stage, if acting via DNA metabolism, should be ineffective in eliciting the typical elongation response. However, it was found that the hormone *was* effective and caused considerable growth, thus suggesting a site of action not associated with DNA.

The question nevertheless is still open and the general opinion is that while GA can exert its hormonal action in a manner not connected with DNA, final consummation of the response necessitates a region in which active cell multiplication is taking place or which in some short time previous to the GA's appearance has taken place.

Applying physico-chemical techniques, Kessler and Snir[8] have indicated that *in vitro* there is a possibility that GA_7 attaches to DNA extracted from higher plants, such as beans, peas, cucumbers and spinach and that it may be "intercalated" adjacent to the adenine–thymine base pairs in the DNA. This intercalation is similar to that of actinomycin's into DNA as mentioned in Chapter 8. It is claimed that as a result of GA insertion into the strand, certain changes, possibly the formation of loops, are caused in the double helix and thereby cause a change in the "functional template". It is of interest that GA_3 which differs from GA_7 in only one hydroxyl group (see Fig. 43) was not inserted into any of the DNA species tested. These experiments suggest that at least some of the gibberellins may act in this manner. However, since results are from *in vitro* conditions and only pertain to one of the wide array of gibberellins more evidence is needed to show that *in vivo* a similar possibility exists.

A further elaboration of the possibility that GA may act at the gene level is that the hormone may induce synthesis of specific mRNA molecules for enzymes participating in growth. A system which has been used extensively in this context is the germination of barley grains and this shall be outlined briefly.

In 1960 Yomo in Japan and independently Paleg in Australia, who worked on beer fermentation, reported that GA triggers the appearance of α-amylase in the storage tissue or endosperm of the germinating grain. This enzyme hydrolyses starch to reducing sugar and in practice it is possible to determine the amount of GA in the seed or in any applied extract by direct measurement of reducing sugar in the system. It was subsequently found that the GA is formed in the scutellum and the embryo from whence it migrates to the *aleurone* tissue—this being a thin layer (often single in wheat and triple in barley) of living cells encompassing the starchy endosperm. Responding to the GA stimulus the aleurone activates or causes a *de novo* synthesis of a series of enzymes including α-amylase, proteases, RNases, pentosanases, etc. (see review by van Overbeek[15]).

Varner and Johri[16] executed a series of experiments which indicated that RNA is involved in the above process. Their fundamental data, gathered from trials conducted in aleurone layers isolated from the rest of the seed to which GA, starch, inhibitors, antimetabolites and labelled material were added, indicated that:

   (i) in order to obtain continual synthesis of RNA and α-amylase the presence of GA is required throughout and its function does not cease with the commencement of RNA synthesis (see also discussion on *switch* and *relay* on p. 143 of Chapter 16);

  (ii) it is possible to inhibit enzyme synthesis by application of RNA synthesis inhibitors.

The data presented in Table 7 shows that the formation of α-amylase by the isolated aleurone layers is sensitive to actinomycin D only during the first few hours after addition of gibberellic acid. Actinomycin added 7 hours after the addition of GA has little effect. However, fluorophenylalanine added at this time is still effective.

These results are consistent with the postulate that GA induces the formation of a specific mRNA which directs the *de novo* synthesis of the enzyme. Within a few hours after the addition of GA the quantity of mRNA is no longer rate-limiting in α-amylase synthesis. From this time on, the enzyme's synthesis would still be susceptible to protein synthesis inhibitors.

TABLE 7. SENSITIVITY OF α-AMYLASE SYNTHESIS TO INHIBITORS
(According to Varner and Ram Chandra[17])

Each sample contained ten isolated aleurone layers in the incubation medium as mentioned in the text. (The actinomycin and the fluorophenyl alanine were added either at the same time as the GA or after 7 hours. Total incubation period was 30 hours.)

Treatment	μg amylase
Control—no GA	13
GA_3 10^{-6} M	66
GA_3 + actinomycin D 100 μg/ml	24
GA_3 + actinomycin (after 7 hours)	55
GA_3 + fluorophenyl alanine	12
GA_3 + fluorophenyl alanine (after 7 hours)	12

Support for this assumption is perhaps provided by Zolotov and Leshem[19] who extracted RNA from embryos of germinating barley and applied it to a system containing aleurones from embryoless endosperm halves, starch and buffer medium. As seen in Fig. 44 the applied RNA induced partial hydrolysis of the starch—an effect similar to that produced by the hormone itself.

Another way in which the above assumptions could be tested would be the determination of interaction between GA and isolated nuclei of aleurone cells. Since the isolation of nuclei from aleurone cells presents several technical difficulties, isolated nuclei of dwarf-pea plants were used instead. Nuclei isolated by Johri and Varner[6] from pea tissue growing in a medium containing GA exhibited a greater capacity for DNA-dependent RNA synthesis than nuclei isolated from tissues not containing the hormone. The RNA synthesized in the nuclei from the GA-treated tissue had a higher molecular weight and possessed a different base sequence from the RNA from control nuclei. Upon fractionation of the hormone-induced RNA on a MAK column it was found

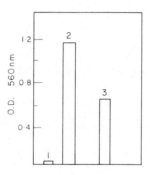

FIG. 44. Production of reducing sugar by isolated barley aleurone layers in starch solutions containing: 1. buffer medium only, 2. buffer medium + RNA isolated from germinating embryos, 3. buffer medium + 25 ppm GA$_3$. (Data of Zolotov and Leshem[19].)

mainly in association with DNA or was located in the so-called "tenaciously bound RNA"† fraction, thus suggesting mRNA. The general conclusion drawn is that the nuclei respond to GA only if they are extracted in its presence: when GA was added at later stages during the process of isolation the above-described response was retarded considerably. It therefore appears that the cytoplasm or the nucleus contains a factor (or factors) which has to be activated before GA exerts its final effect expressed as synthesis of specific RNA's. The inability of nuclei isolated in media lacking the hormone to respond is interpreted as loss of the factor during isolation or alternatively that GA is some-

† RNA adhering to the MAK after elution with an NaCl gradient is termed "tenaciously bound". It may be freed by subsequent elution with hot base or detergent and is believed to have mRNA-like properties.

how responsible for the transfer of the factor from the cytoplasm to the nucleus. The nature of the factor and its connection with GA and RNA synthesis is at present unknown, but it is postulated that some sort of physical interaction exists, e.g. non-covalent binding of the hormone to a macromolecule or to the factor itself. It is also suggested that the factor itself is a protein.

A different approach is taken by Yung and Mann[18] who suggest that GA causes mRNA activation induced by its increased demand. Since demand rises the gene responds by transcribing more messenger. Ram-Chandra and Duynstee[14] have produced some interesting evidence which attributes the function of GA to a methylating process similar to that described for cytokinins on p. 123. They found that the methylation of purines, particularly of the adenosine residues of RNA is favoured in the hormone-treated cells and that in general GA treatment results in changes in the extent, degree and pattern of nucleic acid methylation. Whether this acts as an aid in correct translation has yet to be determined.

Finally, we mention that it also has been claimed that GA, like IAA, may act as a histone derepressor or may be an mRNA stabilizer.

(b) *GA and membranes*

Using the classic barley-endosperm system Paleg *et al.*[12] have presented evidence that at least one basic site of GA action may be the cell membrane. The hormonal action on cell membranes has been mentioned on p. 94 in Chapter 12 and it is recalled that physiological effects of certain mammalian hormones are membrane-associated, and as previously stated, auxin too may affect membranes. Paleg's group, using the electron microscope, has identified lysosome-like organelles in aleurone layers. It is suggested that these units which as lysosomes are sacs of membrane-contained hydrolytic enzymes play an important role in conversion of endosperm reserves to soluble products to be utilized by the developing embryo. GA may exert its action by increasing the permeability of the membranes encompassing these lysosome-like units and as a result cause an increased diffusion of hydrolytic enzymes into the adjacent endosperm tissue. This possibility pertains only to those enzymes which undergo *activation* by GA and does *not* explain the

GA-induced *de novo* synthesis of α-amylase. The latter case may involve another site of GA action and is in accordance with the concept that a given hormone has more than one site or mode of action.

(c) *GA as an auxin inducer*

It has been suggested that the hormonal effect of GA may be an indirect one mediated by IAA. van Overbeek[15] has described such a situation which may be presented as follows:

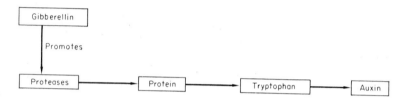

GA activates hydrolysing enzymes, including proteases which in turn digest protein to amino acids including tryptophan which is considered an auxin precursor. By this devious means it is possible for GA to induce auxin production and the latter hormone is considered responsible for the hormonal response finally elicited. Based on this assumption van Overbeek has suggested that the cereal coleoptile, for example, grows as follows. The seed embryo after the imbibition begins to produce GA which migrates to the aleurone, there inducing or activating hydrolytic enzymes including proteases. These act on the protein fraction of the endosperm reserves, amino acids, including tryptophan, are released. This amino acid is transferred to the shootlet and upon reaching the terminal apex is converted into IAA and from this stage onwards the pattern of shoot development is a typically IAA-induced response.

In the germinating cereal grain, as in several other tissues,[5,9] the above mode of auxin promotion is not exclusive, since it has been found in the author's and other laboratories that GA may affect peroxidase activity, which normally increases during the course of germination. It is also believed that the endogenous level of auxin is regulated by this enzyme. IAA peroxidase degrades auxin to methylene-oxyindole and by application of exogenous GA, or under the influence of endogenous GA formed by the embryo, it is possible to regulate the enzyme's activity thus providing a non-protease–tryptophan-linked metabolic pathway of auxin regulation. However, auxin synthesis undoubtedly occurs in the formerly described manner, so van Overbeek's scheme as indicated above may be extended as follows:

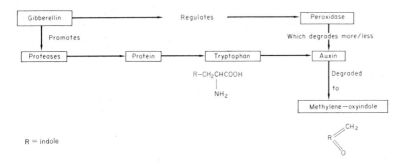

A further mode of gibberellin–auxin interaction is the possibility that GA aids the release of "bound auxin" and thus doing promotes its ability to diffuse from cell to cell. It also has been claimed that GA may act as a histone derepressor.

(d) *GA and cyclic adenosine monophosphate*

The general role of the cyclic nucleotide in molecular processes has been discussed in Chapter 4 and its possible function in mediation of hormonal effects has been outlined on pp. 95–98. Mammalian hormones have been the most extensively studied in this respect, and to date the comparatively few published reports on its interaction with phytohormones deal mainly with GA though some basic information has been provided on auxins (see, for example, Kamisaka and Masuda[7]) and on cytokinins as well. The involvement of the cAMP system in the latter case is discussed under the section "Cytokinesins" in Chapter 14.

It is recalled that according to the hypothesis offered by Butcher *et al.* as seen in Fig. 33 on p. 97, the hormone is regarded as the "first messenger" which acts on a membrane-contained adenyl cyclase unit inducing cyclization of ATP to $3'5'$-cAMP, which is termed the "second messenger" or "deputy". This in turn stimulates protein kinases which are a group of phosphorylating enzymes which may activate inactive enzyme precursors and thus elicit physiological effects.

Some of the initial descriptive work in plant systems was performed by Galsky and Lippincot,[3] Nickells *et al.*[10] and Earle and Galsky.[2] Results (Table 8) of a typical experiment utilizing the barley endosperm system indicated that cAMP promoted the production of α-amylase in the absence of GA, and enhanced production when GA was present in low concentrations. When GA was applied at a high concentration, 2·6 nM, the nucleotide was inhibitory. Other experiments showed that in addition to the cAMP, ADP was also capable of producing similar effects.

TABLE 8. EFFECT OF CYCLIC AMP AND GA_3 ON α-AMYLASE SYNTHESIS IN BARLEY ENDOSPERM HALVES

(Data of Galsky and Lippincot[3])

Treatment	Water		GA, 0·26 mμM		GA, 0·78 mμM		GA, 2·60 mμM	
	−	+	−	+	−	+	−	+
Units enzyme activity	1·0	2·3	1·4	2·6	2·0	3·0	3·3	2·7
Per cent change as compared to cAMP lacking control		+130		+86		+67		(−)

Two alternative interpretations are given to the promotion of α-amylase by the two nucleotides investigated:

(i) they could induce GA synthesis;
(ii) they could possibly "mimic" the effect of GA by activating the same processes that the hormone does.

It is claimed that the first possibility does not apply since the addition of AMO-1618, an inhibitor of gibberellin synthesis, had no deleterious effect on the production of α-amylase in the presence of either of the two nucleotides. Had the nucleotide effect been mediated by GA the AMO-1618 should have inhibited the appearance of the enzyme since, as previously outlined, this enzyme appears *de novo* as a result of GA action on aleurone layers. This argumentation therefore suggests that the second alternative, viz. nucleotide mimicry of GA, applies in the present case.

Interpretation of cAMP action has mainly sought analogy to the membrane-contained cyclase hypothesis. Another site of action overlooked at least in interpretation of observed effects is the possible interaction with DNA either by combination with polymerase or CAP protein as suggested by Pastan and Perlman (see Fig. 13 in Chapter 4).

The above and other communications, while convincingly indicating that cAMP mimics GA action, prove neither the occurrence of cAMP nor its metabolism in higher plants. Pollard[13] has presented evidence that may indicate that the key enzyme adenyl cyclase is present in barley aleurones since labelled adenine was converted to cAMP. Moreover, the cAMP isolated from GA-treated tissues contained twice as much the amount of radioactivity as that from control. Finally, it is mentioned that while cAMP metabolism, mode of action, presence, etc., are extensively detailed in mammals, the cyclic nucleotide nevertheless was initially observed (and termed *acrasin*) in lower plants—the Myxomycetes—where it functions in the process of aggregation.

RECOMMENDED READING AND REFERENCES

1. DEGANI, Y., ATSMON, D. and HALEVY, A. H., DNA synthesis and hormone induced elongation in the cucumber hypocotyl. *Nature*, **288**, 554–7 (1970). An interesting effect of GA leading to doubts as to involvement of nuclear DNA.
2. EARLE, K. M. and GALSKY, A. G., The action of cyclic AMP on GA_3-controlled responses. II. Similarities in the induction of barley endosperm ATPase activity by gibberellic acid and cyclic 3'5'-adenosine monophosphate. *Pl. Cell Physiol.*, **12**, 727–33. See no. 3 below.
3. GALSKY, A. G. and LIPPINCOT, J. A., Promotion and inhibition of α-amylase production in barley endosperm by cyclic 3,5,AMP and ADP. *Pl. Cell Physiol.*, **10**, 609–20 (1969). Together with nos. 2 and 10 descriptive work establishing effect of the cyclic nucleotide on plant enzyme systems. Indication that it does not induce GA synthesis but rather mimics its effect.
4. HABER, A., FOARD, D. E. and PERDUE, S. W., Actions of gibberellic and abscisic acids on lettuce seed germination without actions on nuclear DNA synthesis. *Pl. Physiol.*, **44**, 463–7 (1969). An experiment which differentiates between GA-promoted growth and DNA metabolism.
5. HALEVI, A. H., Interaction of growth-retarding compounds and gibberellin on indoleacetic acid oxidase and peroxidase of cucumber seedlings. *Pl. Physiol.*, **38**, 731–7. See no. 9 below.
6. JOHRI, M. M. and VARNER, J., Enhancement of RNA synthesis in isolated pea nuclei by gibberellic acid. *Proc. Nat. Acad. Sci.*, **59**, 269–76 (1968). See no. 17 below.

7. KAMISAKA, S. and MASUDA, Y., Stimulation of auxin induced cell expansion in plant tissue by cyclic 3'5' adenosine mono-phosphate. *Naturwissenschaften*, **57**, 546–54 (1970). The effect of the cAMP outlined in connection with auxin as well.

8. KESSLER, B. and SNIR, I., Interactions *in vitro* between gibberellins and DNA. *Biochim. Biophys. Acta*, **95**, 207–18 (1969). Elegant experiments indicating that GA may intercalate between certain bases of DNA extracted from higher plants.

9. McCUNE, D. C. and GALSTON, A. W., Inverse effects of gibberellin in peroxidase activity and growth in dwarf strains of peas and corn. *Pl. Physiol.* **34**, 416–18 (1959). An alternative mode of GA action—GA may regulate the enzyme which controls endogenous IAA levels.

10. NICKELLS, M. W., SCHAEFER, G. M. and GALSKY, A. G., The action of cyclic-AMP on GA_3-controlled responses. I. Induction of barley endosperm protease and acid phosphatase activity by cyclic 3'-5'-adenosine monophosphate. *Pl. Cell Physiol.*, **12**, 717–25 (1971). See no. 3 above.

11. NITSAN, Y. and LANG, A., DNA synthesis in the elongating non-dividing cells of the lentil epicotyl and its promotion by gibberellin. *Pl. Physiol.*, **41**, 965–70 (1966). A pioneering effort which blazed the trail for investigation of DNA involvement in GA-elicited physiological responses.

12. PALEG, L., WOOD, A. and SAWNHEY, R. K., Control of permeability by Gibberellin, in *Plant Growth Substances*, 1970, Proc. Int. Cong. on Growth Substances, Canberra, 1971. Abstracts. A suggestion, backed with experimental evidence, that in the aleurone GA may release certain enzymes by increasing permeability of membranes.

13. POLLARD, C. J., Influence of gibberellic acid on the incorporation of $8-^{14}C$ adenine into adenosine 3'-5'-cyclic phosphate in barley aleurone layers. *Biochim. Biophys. Acta*, **201**, 511–12 (1970). Indirect evidence that adenyl cyclase is present in plants.

14. RAM-CHANDRA, G. and DUYNSTEE, E. E., The involvement of nucleic acid methylation in the gibberellic acid evoked enzyme synthesis in aleurone layers, in *Plant Growth Substances*, 1970, Proc. Int. Cong. on Growth Substances, Canberra, 1971. Abstracts. A methylating effect of GA recalling a like one suggested for cytokinins.

15. VAN OVERBEEK, J., Plant hormones and regulators. *Science*, **152**, 721–31 (1966). An interesting overall review on modes of action of the major classes of phytohormones.

16. VARNER, J. E. and JOHRI, M. M., Hormonal control of enzyme synthesis, in *Biochemistry and Physiology of Plant Growth Substances*, pp. 793–814, ed. P. WIGHTMAN and G. SETTERFIELD, Runge Press, Ottawa, 1968. See no. 17 below.

17. VARNER, J. E. and RAM-CHANDRA, G., Hormonal control of enzyme synthesis in barley endosperm. *Proc. Nat. Acad. Sci.*, **52**, 100–6 (1964). This research, together with nos. 4 and 16, provide biochemical proof that some of the GA effects are mediated by RNA.

18. YUNG, K. A. and MANN, J. D., Inhibition of early steps in the gibberellin activated synthesis of α-amylase. *Pl. Physiol.*, **42**, 195–200 (1967). A different approach to the problem. GA may act by enhancing *demand* for mRNA thereby increasing its *supply*.

19. ZOLOTOV, A. and LESHEM, Y. (1968). See reference no. 14 to Chapter 10.

CHAPTER 16

Mode of Action of Ethylene and Abscisic Acid

IT HAS now been established that two further substances, the gas ethylene C_2H_4 and abscisic acid (see Fig. 45), are important endogenous plant-growth regulators.

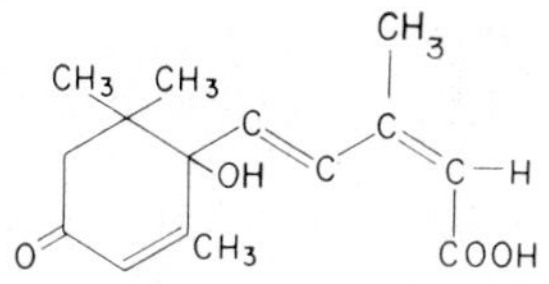

FIG. 45. Abscisic acid.

Since both of these substances are often associated with overall inhibition or abnormal growth the question has been raised if these may be considered true phytohormones. The fact that they are classified in this category is based on the logic that differentiation and growth are a function of equilibrium between promotive and inhibitory processes and that many phenomena involving cessation of development, e.g. leaf abscission, onset of seed and bud dormancy, prevention of excessive lateral branching by apical dominance, etc., are as essential to the survival of the plant as positive growth. The endogenous growth regulators such as abscisic acid and ethylene which, together with other growth-regulating substances, contribute to these processes may therefore be considered hormones and this is all the more so since they fulfil other criteria for qualification as hormones. It has also been demonstrated that not all abscisic acid or ethylene-associated effects are inhibitory and that flowering or elongation, for example, of certain plants may

141

be *promoted* by them thus showing that these are not necessarily endogenous inhibitors.

Ethylene

It has long been known that C_2H_4 is given off by ripening fruit and the gas has been used to promote maturation of green produce. It has also been shown that many effects that were attributed to auxin are actually induced by ethylene and it has been claimed (Burg and Burg[3]) that several auxin-controlled processes necessitate the auxin's "deputy" which is ethylene and in such cases the gas is a mediator between the auxin and its final effect. According to this hypothesis we here encounter *feedback-control* of auxin translocation and synthesis. Much of the work performed to date on ethylene is of a descriptive nature but what appears quite evident is that auxin induces ethylene production which upon becoming supraoptimal may shut off auxin synthesis (or translocation) thereby causing a decrease in ethylene production which permits renewal of auxin synthesis.

In their review Burg *et al.*[2] contend that ethylene binds to a metal-containing receptor having a specificity typical of a protein. This binding requires O_2 and is competitively inhibited by CO_2. The ethylene-receptor link is apparently non-covalent for it is readily reversible. Concerning the possible identity of the receptor protein, an interesting postulation has been made that it may be associated with the hydroxyproline-rich wall protein "extensin" mentioned on p. 111. It is recalled that this is a glycoprotein potentially capable of extensive cross-linking. Ridge and Osborne[9] have demonstrated that ethylene increases cytoplasmic hydroxylation of proline, leading to enrichment in specific hydroxy-proline-rich wall peroxidase, and it is known that maturation and loss of wall extensibility are connected with high hydroxyproline levels. While experimental evidence referred to wall peroxidases the possibility was suggested that other proteins, such as extensin, are also hydroxylated.

The site of ethylene action is not only the cell wall but may also be associated with the cell membrane and it has been indicated that the gas increases passive permeability since it is a highly fat-soluble molecule which should concentrate in lipid phases such as membranes.

Against this it is known that ethylene is also appreciably soluble in water as well and would not be expected to seek out lipid phases.

Be the cell wall or membrane the receptor site or not, Burg *et al.* state that an immediate result of ethylene binding is inhibition of polar auxin transport and cell division but *promotion* of radial cell expansion.

Secondary effects of ethylene action include alteration of microtubule structure, microfibril orientation and also DNA metabolism. The latter is still a moot point, that is, whether this is a primary or a secondary effect, but in either event several enzymes are markedly affected by ethylene, e.g. mitochondria-contained ATPase activity is enhanced, but it has not yet been determined if this is an effect associated with either mitochondrial or nuclear nucleic acid metabolism or if activation is by some other means.

Another key enzyme affected by ethylene is *phenyl-ammonia-lyase* (PAL) which participates in plant flavonoid synthesis and whose production and activity have been related to phytochrome. Riov *et al.*[10] have established that the continuation of synthesis of PAL in citrus-peel tissue necessitates a steady supply of ethylene, and if the supply is cut off PAL ceases to be synthesized. This illustrates a fundamental principle in ethylene—or any other phytohormone—effect since an obvious question is raised: Does the hormone function only as a process switch or a *trigger*, or, in addition, is it necessary throughout? In the latter case we speak of necessity of hormone for *relay* (a term taken from electricity) of information, and once the trigger is removed relay ceases. In this respect no general rule has been provided for all phytohormones, but it appears that in the case of ethylene and of certain cytokinin-induced phenomena the hormone is essential for both triggering and relay.

Abscisic acid

Abscisic acid (ABA) is an isoprenoid in structure and may possibly be the derivative of the carotene compound violaxanthin, which somewhat resembles vitamin A in structure and is synthesized from mevalonate. The hormone participates in many processes including dormancy induction, senescence and formation of abscission layers, and hence its name. It is also known that ABA may inhibit flowering of certain long-day plants and promote flowering of some short-day plants and is connected with ability to withstand moisture stress.

Compared to the other phytohormones, IAA, GA and cytokinins, comparatively sparse information exists on the possible mode of action of ABA, but what appears to be evident is that it may in certain cases perform the role of a GA antagonist, and it has recently been documented that besides GA other hormones too may be affected by ABA. Pearson and Wareing[8] have shown that ABA has a marked inhibitory effect on plant chromatin and it is known that it inhibits the synthesis of DNA and of all species of RNA—ribosomal RNA appearing the most sensitive and the sRNA fraction possibly less so. It has also been suggested that ABA may have a specific inhibitory effect on mRNA coding for certain protein enzymes and may also prevent the association of pre-existing mRNA's with ribosomes to form protein-synthesizing polysomal units. In germinating *Fraxinus* seeds Villiers[13] has reported that ABA functions as a dormancy inducer by the above mechanism. In this case it would be of interest to ascertain whether ABA action is direct and acts on one of the structural components involved, or whether the actual site of inhibition may be on *initiation* or *elongation factors* which, as mentioned in Chapter 6, are essential for protein synthesis.

Concerning the RNA decrease reported by various laboratories two alternative explanations, not necessarily mutually exclusive, may be offered. The first and more commonly believed is that ABA may have inhibitory action on the enzyme RNA polymerase and the transcriptional block thus formed prevents synthesis of DNA-primed RNA. Rather interesting evidence supporting this approach has been provided by Bex[1] who detected a marked decrease in the specific activity of RNA polymerase when maize coleoptiles were incubated with ABA. An alternative explanation suggested by Leshem[5] states that the same overall effect, viz. decrease in RNA levels, could be obtained by ABA *activation* of the enzyme RNase and not necessarily by *inhibition* of RNA polymerase. The mode of RNase activation is elusive and appears not to be a direct allosteric effect as could possibly be expected.[6]

ABA and allostery. From the results of investigation of kinetics of action of the enzyme invertase, which hydrolyses sucrose to monosaccharides, and its interaction with ABA and benzyladenine of the cytokinin group, van Overbeek *et al.*[12] and Saunders and Poulson[11] have suggested that ABA may serve as a negative allosteric effector on biologically active proteins. The observation of Glinka and Reinhold[4] that ABA profoundly affects the permeability of all membranes to passage of water is another aspect of the hormone's action. This observation possibly provides a cogent explanation for the now well established fact, of considerable practical importance, that ABA treatment imparts the ability to withstand water stress. However, the above suggest that ABA may well have allosteric effects on other biologically active proteins which may include enzymes participating in GA biosynthesis, and on DNA polymerase as possibly indicated by the observed inhibition of DNA synthesis caused by the hormone.

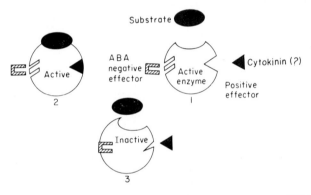

Fig. 46. The allosteric effect of ABA. (After van Overbeek *et al*[11].)

Van Overbeek *et al.*[12] have proposed an overall scheme for ABA action as a *negative* allosteric effector on biologically active proteins (see Fig. 46). The *positive* effector or *activator* in their model may be more specific and it is postulated that in a certain system, for example, it may be GA while in another cytokinin or auxin. The biologically active protein itself may be any one of the previously mentioned enzymes or some other functional protein.

Milborrow[7] has suggested that one of the enzymes qualifying for the ABA-effected allosteric protein is RNA polymerase since a series of experiments indicated that ABA prevented incorporation of labelled uridine and thymidine into nucleic acids, while no inhibition of amino-acid incorporation into protein was experienced. From this it is deduced that ABA does not act at a translational level but, as mentioned above, rather at a transcriptional one. It is postulated that ABA attachment to the protein is by means of its two polar groups which link to specific "allosteric sites" situated on the protein (stage 3 in Fig. 46). The attachment of ABA causes a steric change of the "substrate site", preventing the entry of the substrate and thus inhibiting biological activity. This may explain why under certain experimental circumstances ABA prevents the expected growth promotion in the presence of benzyl-adenine. If, however, the activator, i.e. the promoting hormone, is attached beforehand (stage 2 in Fig. 46) it lends stability to the protein, prevents deformation of the "substrate site" and attachment of the ABA, thus enabling biological activity to proceed normally. This model outlines how *in vivo* growth may be the function of equilibrium attained between promotory and inhibitory factors.

RECOMMENDED READING AND REFERENCES

1. BEX, J. H. M., Effects of abscisic acid on the soluble RNA polymerase activity in maize coleoptiles. *Planta*, **103**, 11–17 (1972). Experimentation indicating that levels of RNA may be decreased by ABA by means of RNA polymerase inhibition.

2. BURG, S., APELBAUM, A., EISINGER, W. and KANG, B. G., Physiology and mode of action of ethylene. *Hortscience*, **6**, 359–64 (1971). A concise review of mode of ethylene action written by an active group of workers on the subject.

3. BURG, S. P. and BURG, E. A., Auxin stimulated ethylene formation, its relationship to auxin inhibited growth, root geotropism and other plant processes, in *Biochemistry and Physiology of Plant Growth Substances*, ed. P. WIGHTMAN and E. SETTERFIELD, Runge Press, Ottawa, 1968. A reappraisal of several phenomena formerly regarded as typical auxin-elicited effects.

4. GLINKA, Z. and REINHOLD, L., Abscisic acid raises the permeability of plant cells to water. *Plant Physiol.* **48**, 103–5 (1971). A basic observation providing the link between the hormone and water balance in plants.

5. LESHEM, Y., Abscisic acid as a ribonuclease promotor. *Physiologia Pl.*, **24**, 85–89 (1971). ABA may possibly promote RNase and not necessarily inhibit RNA polymerase.

6. LESHEM, Y. and SCHWARZ, L., The selective effect of abscisic acid on ribonucleic acid components. *Physiol. Pl.* **26**, 328–31 (1972). Ribosomal RNA seems more prone to ABA inhibition than the sRNA fraction.

7. MILBORROW, B. V., The occurrence and function of abscisic acid in plants. *Sci. Prog., Oxf.*, **57**, 533–58 (1969). An excellent survey of recent experimentation and its theoretical implications.

8. PEARSON, J. A. and WAREING, P. F., Effect of abscisic acid on activity of chromatin. *Nature* **221,** 672–3 (1969). Chromatin-inhibition caused by ABA suggesting involvement of DNA polymerase.

9. RIDGE, I. and OSBORNE, D., Role of peroxidase when hydroxyproline-rich protein in plant cell walls is increased by ethylene. *Nature, N.B.,* **229,** 205–8 (1971). Ethylene metabolism is linked with changes induced in specific cell wall components.

10. RIOV, J., MONSELISE, S. P. and KAHAN, R. S., Ethylene controlled induction of phenylalanine ammonia lyase in citrus fruit peel. *Pl. Physiol.* **44,** 631–5 (1969). An experimental demonstration of the *relay* phenomenon in hormonal action.

11. SAUNDERS, P. F. and POULSON, R. H., Biochemical studies on the possible mode of action of abscisic acid, an apparent allosteric inhibition of invertase activity, pp. 1581–91 in book as mentioned in no. 3 above. Application of the principle of "allostery" to ABA action.

12. VAN OVERBEEK, J., LOEFFLER, J. E. and MASON, M. I., Dormin (Abscisin II) inhibitor of plant DNA synthesis? *Science,* **156,** 1497–9 (1967). See no. 10 above.

13. VILLIERS, T. A., An autoradiographic study of the effect of the plant hormone abscisic acid on nucleic acid and protein metabolism. *Planta,* **82,** 342–54 (1968). An interesting electron microscope-aided observation that dormancy in *Fraxinus* may be caused by ABA-induced prevention of polysome assembly.

Author Index

149

Subject Index